REXS 2013 – Workshop on Resonant Elastic X-ray Scattering in Condensed Matter

Journal of Physics: Conference Series Volume 519

Oxford, United Kingdom
15-19 July 2013

ISBN: 978-1-63439-300-3
ISSN: 1742-6588

Printed from e-media with permission by:

Curran Associates, Inc.
57 Morehouse Lane
Red Hook, NY 12571

Some format issues inherent in the e-media version may also appear in this print version.

Printed by Curran Associates, Inc. (2014)

For permission requests, please contact the Institute of Physics
at the address below.

Institute of Physics
Dirac House, Temple Back
Bristol BS1 6BE UK

Phone: 44 1 17 929 7481
Fax: 44 1 17 920 0979

techtracking@iop.org

Additional copies of this publication are available from:

Curran Associates, Inc.
57 Morehouse Lane
Red Hook, NY 12571 USA
Phone: 845-758-0400
Fax: 845-758-2634
Email: curran@proceedings.com
Web: www.proceedings.com

TABLE OF CONTENTS

012001 BASICS OF RESONANT ELASTIC X-RAY SCATTERING THEORY 1
S Grenier, Y Joly

012002 RESONANT ELASTIC X-RAY SCATTERING IN LIFE SCIENCE, CHEMISTRY AND MATERIALS SCIENCE; RECENT DEVELOPMENTS 8
J R Helliwell

012003 AN X-RAY STUDY OF THE DZYALOSHINSKII-MORIYA INTERACTION IN THE WEAK FERROMAGNET $FEBO_3$ 20
V E Dmitrienko, E N Ovchinnikova, S P Collins, G Nisbet, G Beutier

012004 MAGNETIC PHASE DIAGRAM AND ORDERED GROUND STATE OF $GDMN_2O_5$ MULTIFERROIC STUDIED BY X-RAY MAGNETIC SCATTERING 26
C Vecchini, A Bombardi, L C Chapon, N Lee, P G Radaelli, S-W Cheong

012005 EXISTENCE OF COVALENT-LIKE BONDING IN $PD_{40}CU_{20}GE_{40}$ METALLIC GLASS OBSERVED BY ANOMALOUS X-RAY SCATTERING 35
T Yamamoto, S Hosokawa, J-F Bérar, N Boudet, W-C Pilgrim, K Hayashi, H Kato

012006 PHASE TRANSITION OF KDP OBSERVED BY RESONANT X-RAY DIFFRACTION AT FORBIDDEN REFLECTIONS 39
G Beutier, S P Collins, E N Ovchinnikova, G Nisbet, V E Dmitrienko

012007 MAGNETIC ORDER IN $GDMNO_3$ IN MAGNETIC FIELDS 49
A Skaugen, D K Shukla, R Feyerherm, E Dudzik, Z Islam, J Strempfer

012008 RESONANT X-RAY SCATTERING IN $LA_{1-X}SR_{1+X}MN0_4$ ($X\geq0.5$): INCOMMENSURATE-LATTICE MODULATION VS. CHARGE-STRIPE MODELS 53
G Subías, J García, J Herrero-Martín, J Blasco, M C Sánchez

012009 RESONANT X-RAY SCATTERING FROM CHIRAL MATERIALS: A-QUARTZ, A-BERLINITE, AND TELLURIUM 59
M Takahashi, J-I Igarashi

012010 COHERENT MAGNETIC DIFFRACTION FROM THE URANIUM M_4 EDGE IN THE MULTI-K MAGNET, USB 66
J A Lim, E Blackburn, G Beutier, F Livet, N Magnani, A Bombardi, R Caciuffo, G H Lander

012011 EVALUATION OF STRUCTURE MODELS OF HO_2PDSI_3 USING DAFS, INTER ALIA AT A SATELLITE REFLECTION 74
M Nentwich, M Zschornak, C Richter, D C Meyer

012012 A NEW CHIRAL PHASE OF $BIFEO_3$ EVIDENCED FROM RESONANT X-RAY BRAGG DIFFRACTION 80
A Rodriguez-Fernandez, S W Lovesey, S P Collins, G Nisbet, J A Blanco

Author Index

Journal of Physics: Conference Series **519** (2014) 011001 doi:10.1088/1742-6596/519/1/011001

Preface

Proceedings of REXS 2013 – Workshop on Resonant Elastic X-ray Scattering in Condensed Matter

15–19 July 2013, Oxford and Diamond Light Source, England

http://www.rexs2013.org/

The aim of this workshop was to bring together experts in experimental and theoretical aspects of resonant elastic x-ray scattering, along with researchers who are new to the field, to discuss important recent results and the fundamentals of the technique. The meeting was a great success, with the first day dedicated to students and new researchers in the field, who received introductory lectures and tutorials. All conference delegates were invited either to make an oral presentation or to present a poster, accompanied by a short talk.

The first two papers selected for the REXS13 proceedings (Grenier & Joly and Helliwell) give a basic background to the theory of REXS and applications across a wide range of scientific areas. The remainder of the papers report on some of the latest scientific results obtained by applying the REXS technique to contemporary problems in condensed matter, materials and x-ray physics. It is hoped that these proceedings provide a snapshot of the current status of a vibrant and diverse scientific technique that will be of value not just to those who attended the workshop but also to any other reader with an interest in the subject.

Local Scientific Committee

REXS13 International Scientific Advisory Committee

M Altarelli, European XFEL, Germany
F de Bergevin, European Synchrotron Radiation Facility, France
J Garcia-Ruiz, Universidad de Zaragoza, Spain
A I Goldman, Iowa State University, USA
M Goldmann, Institut Nanosciences, France
T Schulli, European Synchrotron Radiation Facility, France
C R Natoli, Laboratori Nazionali de Frascati, Italy
G Materlik, Diamond Light Source, UK
L Paolasini, European Synchrotron Radiation Facility, France
U Staub, Paul Scherrer Institut, Switzerland
K Finkelstein, Cornell University, USA
Y Murakami, Photon Factory, Japan

REXS13 Local Scientific Committee

G Beutier, CNRS Grenoble, France
C Mazzoli, Politecnico di Milano, Italy
F Yakhou, European Synchrotron Radiation Facility, France
S D Brown, XMaS UK CRG, France
A Bombardi, Diamond Light Source, UK
S P Collins, Diamond Light Source, UK

Journal of Physics: Conference Series **519** (2014) 011002 doi:10.1088/1742-6596/519/1/011002

Peer review statement

All papers published in this volume of *Journal of Physics: Conference Series* have been peer reviewed through processes administered by the proceedings Editors. Reviews were conducted by expert referees to the professional and scientific standards expected of a proceedings journal published by IOP Publishing.

Journal of Physics: Conference Series **519** (2014) 012001 doi:10.1088/1742-6596/519/1/012001

Basics of Resonant Elastic X-ray Scattering theory

S Grenier and Y Joly

Institut Néel, CNRS & Université Grenoble Alpes, F-38042 Grenoble, France

E-mail: stephane.grenier@grenoble.cnrs.fr

Abstract. We recall the basic equations of Resonant Elastic X-ray Scattering (REXS), starting from the photon-electron Hamiltonian interaction, and describing the different components for the possible transitions. We also make the connection with classical physics and we emphasize the signs of the relevant terms, resonant and non resonant, to comply to the convention on the phase wave.

1. Introduction

The variation with energy of the diffracted peak intensities around the absorption edges has been known since the 1920's, and since the first measurement of a complete spectrum, by Y. Cauchois, in 1956 [1], several works have paved the way to a good understanding of most of the important phenomena, like the studies of D. and L. Templeton on the polarization dependence [2], of V. E. Dmitrienko [3] and of K. Finkelstein[4] on forbidden reflections, of Gibbs [5], Hannon [6] and co-workers on magnetic resonant scattering, or on processes involving non simple dipole transitions again by Templeton and Templeton [7]. The current effort is on making REXS a reliable tool to extract quantitative and valuable informations from the experiments. Nowadays, data are recorded with far better resolution, polarization analysis and with many kinds of sample environments. Theory has also greatly improved even if the description of multi electronic phenomena remains a challenging task.

The purpose of the present paper is to give a brief overview of the main elements of this spectroscopy giving the key points toward the basic equations governing it. We have noticed, that some confusion takes place for example on the convention on the signs in many formulas. Different conventions are possible but very often they cannot be mixed as often done. In simple cases that does not matter, but nowadays one tries to extract tiny details from data from interference terms. Incoherent definitions or conventions become source of errors. We propose thus here not going in detail in the demonstration but insist on the units, the sign and at the end propose, modestly, a general agreement on the conventions to prefer.

Our paper is organized as follows: section 2 gives the main points to describe the interaction of polarized light with matter. From this, the formula for the resonant scattering form factor is demonstrated in section 3. The non resonant terms, Thomson and magnetic scatterings form factors are given in section 4. In section 5 we briefly recall the resonant x-ray scattering theory in the framework of classical physics. Finally, we summarize in section 6 a general convention on the notations and signs for this spectroscopy.

Journal of Physics: Conference Series **519** (2014) 012001 doi:10.1088/1742-6596/519/1/012001

2. Interaction of Polarized light with matter

More general and detailed lectures exist on the interaction of polarized light with matter in text books [8] or in the context of x-ray spectroscopy [9]. Here we focus on the elastic photon in - photon out case. Related by Maxwell equations, and using the second quantization, the **k** components of the vector potential **A**, the electric field **E** and the magnetic field **B** of a propagating plane wave are given by:

$$\begin{aligned} \mathbf{A}(\mathbf{r},t) &= \mathcal{A}_0\left(ae^{i(\mathbf{k}\cdot\mathbf{r}-\omega t)}\epsilon + a^+e^{-i(\mathbf{k}\cdot\mathbf{r}-\omega t)}\epsilon^*\right), \\ \mathbf{E}(\mathbf{r},t) &= -i\omega\mathcal{A}_0\left(ae^{i(\mathbf{k}\cdot\mathbf{r}-\omega t)}\epsilon - a^+e^{-i(\mathbf{k}\cdot\mathbf{r}-\omega t)}\epsilon^*\right), \\ \mathbf{B}(\mathbf{r},t) &= i\mathcal{A}_0\left(ae^{i(\mathbf{k}\cdot\mathbf{r}-\omega t)}\;\mathbf{k}\times\epsilon - a^+e^{-i(\mathbf{k}\cdot\mathbf{r}-\omega t)}\;\mathbf{k}\times\epsilon^*\right), \end{aligned} \tag{1}$$

where a and a^+ are the annihilation and creation operators of the photon, ϵ and $\mathbf{k}$ are its polarization and wave vector and $\mathcal{A}_0$ an amplitude which is determined by imposing the energy of the photon to be $\hbar\omega$. Here we use the most common convention on the sign in the exponent of the exponential; this point is discussed later.

Relativistic effects are negligible except for the non resonant magnetic terms, they were already considered in the proceedings of the REXS 2011 conference [10]. So, here we consider only the non relativistic Hamiltonian $H = H_0 + H_I$. H_0 is the non interacting part and we focus on the interaction part between the field above and an electron of charge $-e$:

$$H_I = \frac{e}{m}\mathbf{p}\cdot\mathbf{A} + \frac{e}{m}\mathbf{S}\cdot\mathbf{B} + \frac{e^2}{2m}\mathbf{A}^2 . \tag{2}$$

The transition operator T is given by:

$$T \approx H_I + H_I G_0(\mathcal{E}_i) H_I, \tag{3}$$

where we have used the first order approximation replacing the Green function G of the Hamiltonian H with the Green function of H_0. $\mathcal{E}_i$ is the initial state energy. G_0 is given by:

$$G_0(\mathcal{E}_i) = \lim_{\eta\to 0^+} \frac{1}{\mathcal{E}_i - H_0 + i\eta} . \tag{4}$$

Expanding T to the second order in e/m, one gets:

$$T = T_1 + T_{2a} + T_{2b} + \dots \tag{5}$$

with:

$$T_1 = \frac{e}{m}(\mathbf{p}\cdot\mathbf{A} + \mathbf{S}\cdot\mathbf{B}) \tag{6}$$

$$T_{2a} = \frac{e^2}{2m}\mathbf{A}\cdot\mathbf{A} \tag{7}$$

$$T_{2b} = \left(\frac{e}{m}\right)^2 (\mathbf{p}\cdot\mathbf{A} + \mathbf{S}\cdot\mathbf{B})\, G_0(\mathcal{E}_i)\, (\mathbf{p}\cdot\mathbf{A} + \mathbf{S}\cdot\mathbf{B}) . \tag{8}$$

T_1 contains two terms linear in **A** or **B**, it describes single photon processes, like absorption or emission. T_2 reveals two-photon processes, one photon in and one photon out, describing scattering processes. It is split in two terms, T_{2a}, which gives the non resonant Thomson scattering as shown in section 3, and T_{2b} responsible for the anomalous process as seen in section 4. From this, the transition probability per unit time is given by the so-called first golden rule [8] and with the convenient normalizations, one finally gets the scattering form factor by the formula:

$$f = \frac{C_s}{-r_0} \langle \phi_f | T_2 | \phi_i \rangle \ . \tag{9}$$

The factor $C_s = -\frac{1}{4\pi\epsilon_0 c^2 \mathcal{A}_0^2}$ takes into account the photon final state density and the normalization with the incoming flux. The normalization by the classical electron radius $\frac{1}{-r_0}$ scales the result in *number of electrons*, $r_0 = e^2/4\pi\epsilon_0 mc^2$ being the elastic scattering amplitude of the free electron.

3. Non resonant form factor

3.1. Thomson scattering

We consider T_{2a} (eq. 8) inserted in equation 9. One gets the Thomson scattering form factor:

$$f_0 = \epsilon_s^* \cdot \epsilon_i \sum_g \left\langle \varphi_g \left| e^{i(\mathbf{k}_i - \mathbf{k}_s)\cdot \mathbf{r}} \right| \varphi_g \right\rangle = \epsilon_s^* \cdot \epsilon_i \int \rho(\mathbf{r})\, e^{i(\mathbf{k}_i - \mathbf{k}_s)\cdot \mathbf{r}} d^3\mathbf{r} \ , \tag{10}$$

where $\rho(\mathbf{r}) = \sum_g |\varphi_g(\mathbf{r})|^2$ is the atomic electron density, the sum being over all the occupied states. In forward scattering one thus gets $f_0 = \epsilon_s^* \cdot \epsilon_i \, Z$, Z being the atomic number. Note that due to the minus sign in the normalisation in equation 9, the Thomson scattering *form factor*, f_0, is positive, while the Thomson scattering *amplitude*, b_0, related to the form factor by the relation $b_0 = -r_0 f_0$, is negative. In contrast to optical energies, at X-ray energies the anomalous term is small making refraction indices to be slightly less than one.

3.2. Non-resonant magnetic scattering

The non-resonant magnetic scattering was observed in 1981 by de Bergevin and Brunel [14]. The Hamiltonian was given by Blume and Gibbs in [15, 16]. It contains the interaction between the magnetic field of the incoming wave with the spin of the electron. It contains also an angular momentum part coming from the expansion to the first order in $\hbar\omega/(\mathcal{E}_n - \mathcal{E}_g)$ of the anomalous contribution T_{2b}. At high energy one gets a magnetic form factor given by $f_{mag} = -i f_m$, with:

$$f_m = \frac{\hbar\omega}{mc^2} \left(\frac{1}{2}\mathbf{L}(\mathbf{Q}) \cdot \mathbf{a} + \mathbf{S}(\mathbf{Q}) \cdot \mathbf{b} \right) \ , \tag{11}$$

where $\mathbf{a}$ and $\mathbf{b}$ are two vectors depending on the incoming and outgoing wave vectors and polarizations:

$$\mathbf{a} = -2(1 - \mathbf{u}_i \cdot \mathbf{u}_s)(\epsilon_s \times \epsilon_i) + (\mathbf{u}_i - \mathbf{u}_s) \cdot (\epsilon_s \times \epsilon_i)(\mathbf{u}_i - \mathbf{u}_s) \ , \tag{12}$$

$$\mathbf{b} = \epsilon_s \times \epsilon_i - (\mathbf{u}_i \times \epsilon_i)(\mathbf{u}_i \cdot \epsilon_s) + (\mathbf{u}_s \times \epsilon_s)(\mathbf{u}_s \cdot \epsilon_i) - (\mathbf{u}_s \times \epsilon_s) \times (\mathbf{u}_i \times \epsilon_i) \ . \tag{13}$$

where $\mathbf{u} = \mathbf{k}/k$. $\mathbf{S}(\mathbf{Q})$ is the Fourier transform of the spin density, that is of $\rho(r)^{\uparrow} - \rho(r)^{\downarrow}$. $\mathbf{L}(\mathbf{Q})$ is related to the Fourier transform of the orbital momentum. It is often a good approximation to take $\mathbf{L}(\mathbf{Q})$ and $\mathbf{S}(\mathbf{Q})$ in the same direction. Note that the expresion of $\mathbf{a}$ is slightly different than the one in the Blume and Gibbs paper [16] (A" term in equation 2) due to a typo already reported by Fabrizi and co-workers [18].

The amplitude of the non-resonant magnetic scattering is very small and neutron diffraction is often far more sensitive to measure magnetic ordering. Nevertheless, when studying magnetic material using resonant processes, because this term interferes with the other terms, it can give notable effects on the intensity of some peaks.

Journal of Physics: Conference Series **519** (2014) 012001 doi:10.1088/1742-6596/519/1/012001

4. Resonant x-ray scattering formula

Introducing the expression of the potential vector and the magnetic field given in eq. 2 in eq. 8, one can distinguish two possible processes. In the first one the emission occurs before the absorption, corresponding to a non resonant process. In the second one the absorption occurs before the emission. The former being far smaller than the latter, it will be neglected. The resonant scattering amplitude depends on transitions from an initial state $|\phi_i\rangle = |\varphi_g, (\epsilon_i, k_i)\rangle$ up to intermediate states $|\varphi_n, 0\rangle$ and from this state down to a final state $|\phi_f\rangle = |\varphi_g, (\epsilon_s, k_s)\rangle$. φ_g and φ_n represent the core and excited electron states. (ϵ_i, k_i) and (ϵ_s, k_s) are the incoming and outgoing photons with the same energy $\hbar\omega$ but different polarizations $\epsilon_{i,s}$ and wave vectors $k_{i,s}$. In this situation, where absorption is before emission, the intermediate state has no photon and we have $\mathcal{E}_i = \mathcal{E}_f = \mathcal{E}_g + \hbar\omega$ where $\mathcal{E}_g$ is the energy of the core state. The energy of the intermediate state $\mathcal{E}_n$ is the energy level of the corresponding electron with the core-hole.

Stating different core states, introducing $\sum_n |n\rangle\langle n| = 1$ on the right side of G_0 and using the fact that on the same right side in eq. 8 applies only the annihilation part of $\mathbf{A}$ and $\mathbf{B}$ and on its left side only the creation part one gets:

$$f' - if'' = -\frac{1}{m}\lim_{\eta\to 0^+}\sum_{n,g}\frac{\left\langle \varphi_g \left| \hat{O}_s^* \right| \varphi_n \right\rangle \left\langle \varphi_n \left| \hat{O}_i \right| \varphi_g \right\rangle}{\mathcal{E}_g - \mathcal{E}_n + \hbar\omega + i\eta} \tag{14}$$

where the operator $\hat{O}$ is given by :

$$\hat{O} = (\epsilon \cdot \mathbf{p} + i\,\mathbf{S}\cdot(\mathbf{k}\times\epsilon))\,e^{i\mathbf{k}\cdot\mathbf{r}} \,. \tag{15}$$

At this stage we perform the Taylor expansion of the exponential up to the second order for the potential vector and to the first order for the magnetic field. Using commutator relations one gets:

$$\left\langle \varphi_n \left| \hat{O} \right| \varphi_g \right\rangle = i\frac{m}{\hbar}(\mathcal{E}_n - \mathcal{E}_g)\langle \varphi_n | \hat{o}_{E1} + \hat{o}_{E2} + \hat{o}_{M1} | \varphi_g \rangle \;, \tag{16}$$

with:

$$\hat{o}_{E1} = \epsilon\cdot\mathbf{r}\,, \tag{17}$$

$$\hat{o}_{E2} = \frac{i}{2}(\epsilon\cdot\mathbf{r})\,(\mathbf{k}\cdot\mathbf{r})\,, \tag{18}$$

$$\hat{o}_{M1} = c_m\,(\mathbf{k}\times\epsilon)\cdot(\mathbf{L}+2\mathbf{S})\;, \tag{19}$$

where $c_m = \frac{\hbar}{2m(\mathcal{E}_n - \mathcal{E}_g)}$. Putting all this together one gets:

$$f' - if'' = m\lim_{\eta\to 0^+}\sum_{n,i}\left(\frac{\mathcal{E}_n - \mathcal{E}_g}{\hbar}\right)^2\frac{\langle \varphi_g | \hat{o}_s^* | \varphi_n\rangle\langle \varphi_n | \hat{o}_i | \varphi_g\rangle}{\hbar\omega - (\mathcal{E}_n - \mathcal{E}_g) + i\eta} \;, \tag{20}$$

which is the basic equation in S. I. unit. Note the minus sign in front of f'' to follow the convention on its positive value. Because $\hbar\omega \approx \mathcal{E}_n - \mathcal{E}_g$ and taking into account the fact that most often the intermediate states spread into a continuum one often prefers the integral form:

$$f' - if'' \approx m\omega^2\sum_g\int_{\mathcal{E}_F}^{\infty}\frac{\frac{1}{\pi}\sum_n\langle \varphi_g | \hat{o}_s^* | \bar{\varphi}_n\rangle\langle \bar{\varphi}_n | \hat{o}_i | \varphi_g\rangle}{\hbar\omega - (\mathcal{E} - \mathcal{E}_g) + i\frac{\Gamma}{2}}d\mathcal{E} \;, \tag{21}$$

where $\mathcal{E}_F$ is the Fermi energy and Γ is a phenomenological broadening due to the finite lifetime. $\bar{\varphi}_n$ are now electronic states normalized by the density of states. The summation on n is now only on states at the energy $\mathcal{E}$. The equation above is monoelectronic which is a serious approximation for example at the L_{23} of $3d$ elements. To go deeper in the simulation of REXS spectra, in these cases, one has to use Multiplet formalism or a time-dependent DFT approach [13]. Bethe-Salpeter equations or even configuration techniques, supposed to be more precise, already used for XANES spectroscopy could also be used.

5. Classical X-ray Resonant Scattering

In classical physics, the photon is described by an electromagnetic field that induces an acceleration on a bound electron of charge $-e$ and mass m. The interaction is governed by the electrical force $-eE$ within Newton mechanics, the speed of electron remains sub relativistic, and we neglect the magnetic field:

$$m\ddot{x} = -eE - \eta\dot{x} - \kappa x \tag{22}$$

where $-eE$ is the force experienced by the electron in the electric field, $\eta\dot{x}$ is a damping force that tends to slow the particle, and is opposite in direction to its speed, and κx is a force that retains the particle around its average position; η and κ are both positive. For a field of the form $E_0(z,t) = E_0\ e^{i(kz-\omega t)}$, in the far field approximation [11], this equation leads to:

$$\frac{E_{rad}(r,t)}{E_0(0,t)} \;=\; -r_0\,\frac{\omega^2}{\omega^2-\omega_0^2+i\gamma\omega}\,\frac{e^{ikr}}{r} \tag{23}$$

where $\gamma = \eta/m$ and $\omega_0 = \sqrt{\kappa/m}$, introducing the minus sign so as to keep positive the Thomson form factor (the high ω limit), giving

$$b \;=\; -r_0\,\frac{\omega^2}{\omega^2-\omega_0^2+i\gamma\omega} = -r_0\ f\ , \tag{24}$$

with f the form factor:

$$f \;=\; \frac{\omega^2}{\omega^2-\omega_0^2+i\gamma\omega}\ . \tag{25}$$

Note that a free charge has a negative scattering amplitude which takes importance when dealing with refraction (interference with the incident field) and multiple scattering effects. When the frequency is much lower than the resonance, one finds the well-known atomic, or small particle scattering amplitude, referred to Rayleigh scattering:

$$b \;\approx\; r_0\frac{\omega^2}{\omega_0^2}\ . \tag{26}$$

In this case the scattering amplitude becomes positive, leading to refractive indices greater than one. One can also write expression (25) to single out a constant term and a dispersive term:

$$f \;=\; 1+\frac{\omega_0^2-i\gamma\omega}{\omega^2-\omega_0^2+i\gamma\omega}\ . \tag{27}$$

Close to resonances one can identify the resonant term introduced in the quantum mechanics development, if γ is small compared to ω and $\omega \approx \omega_0$:

$$f \;\approx\; 1+\frac{\omega_0/2}{\omega-\omega_0+i\gamma/2}\ . \tag{28}$$

The effect of polarization, of the direction of the electric field relatively to the scattering direction is taken into account simply by the scalar product of the incident and the outgoing electric fields, which is multiplied to the scattering amplitude.

6. Intensity and convention

6.1. Phase and diffraction vector

The phase of a propagating plane wave of energy $\hbar\omega$ and wave vector $\mathbf{k}$ is written as:

$$e^{i(\mathbf{k}\cdot\mathbf{r}-\omega t)}. \tag{29}$$

This choice [12] for the phase space and time evolution imposes the signs between various quantities and requires great care in all subsequent analytical developments in resonant x-ray scattering theory. Consider the energy $\mathcal{E} = \mathcal{E}_r + i\mathcal{E}_i$ of an excited state whose decay with time goes as $e^{-t/\tau}$ with a lifetime $\tau > 0$. Convention (29) imposes $\mathcal{E}_i$ to be negative, and $\mathcal{E}_i = -\hbar/2\tau = -\Gamma/2$ with the *width* $\Gamma = \hbar/\tau$. Another example deals with the spatial phase shift taken by the photon over a small volume, like a core $1s$ orbital, leading to $e^{i\mathbf{k}\cdot\mathbf{r}} \approx 1 + i\mathbf{k}\cdot\mathbf{r}$ and giving rise to quadrupolar transitions between states of same parity. For instance, it imposes the sign on the process that consists in a dipolar absorption followed by a quadrupolar emission (so called E1E2 [6]).

The sign of the diffraction vector $\mathbf{Q}$ is also chosen, it appears when dealing with spatial phase shifts $(e^{i(\mathbf{k}_f\cdot\mathbf{r})})^* e^{i(\mathbf{k}_i\cdot\mathbf{r})} = e^{-i(\mathbf{k}_f-\mathbf{k}_i)\cdot\mathbf{r}} = e^{-i\mathbf{Q}\cdot\mathbf{r}}$, that is the choice is $\mathbf{Q} = \mathbf{k}_f - \mathbf{k}_i$.

6.2. Scattering amplitude and cross-section

Incident photon initially with wave function e^{ikz} are described far from the scatterer by a spherically diverging wave of the form $\frac{b(\theta)}{r}e^{ikr}$, where $b(\theta)$ is the scattering amplitude:

$$\psi \simeq e^{ikz} + \frac{b(\theta)}{r}e^{ikr}. \tag{30}$$

The probability that the photon passes through the surface $dS = r^2 d\Omega$ is $cr^{-2}|b|^2 dS = c|b|^2 d\Omega$, c being the speed of light. Normalized by the incident photon density, one obtains the scattering cross-section:

$$d\sigma_{sc.} = \; |b|^2 \, d\Omega. \tag{31}$$

The scattering amplitude induces another phase shift when photoelectric transitions toward excited states of some finite lifetime take place during the scattering process. The scattering amplitude is then a complex number. A general result in scattering theory states that the imaginary part of forward scattering amplitude is related to the total cross-section (scattering + absorption) by the so-called optical theorem [11]:

$$\sigma_{\text{tot.}} = \frac{4\pi}{k}\text{Im}\{b(0)\} \tag{32}$$

The imaginary part of the scattering amplitude should consistently be positive. The scattering form factor f, from $b = -r_0 f$, is written $f = f_0 + f' - if''$ where f_0, the free electron (Thomson) form factor which is positive, as well as the imaginary part f'' which usually is obtained from absorption measurements and taken positive to show a rising edge with increasing energy.

The total diffraction intensity is given, up to the second-order in absorption/emission photon processes, by $I(\omega) \propto |F(\omega)|^2$ where the static structure factor F is given by the expression:

$$F(\omega) = \sum_j e^{-i\mathbf{Q}\cdot\mathbf{R}_j}\left(f_{0j} - if_{mj} + f'_j(\omega) - if''_j(\omega)\right). \tag{33}$$

Journal of Physics: Conference Series **519** (2014) 012001 doi:10.1088/1742-6596/519/1/012001

The sum is over all atoms j of the unit cell. Of course, when calculating intensity, the complex conjugate can be taken. This formula does not take into account the dynamical process in diffraction. A step to overcome this approximation permits to simulate the birefringence effect [10]. In this case the rotation of the polarization along the propagation of the field in the sample is taken into account. The scattering by a specific unit cell thus depends on the polarization and the amplitude of the electromagnetic field at this point in the sample.

References

[1] Cauchois Y *Comptes Rendus de l'Académie des Sciences (Paris)* 1956 **242** 100
[2] Templeton D H and Templeton L K 1982 *Acta Cryst.* A **38**, 62
[3] Dmitrienko V E 1983 *Acta Cryst.* A **39**, 29
[4] Finkelstein K D, Qun Shen and Shastri S 1992 *Phys. Rev. Lett.* **69** 1612
[5] Doon Gibbs, Harshman D R, Isaacs E D, McWhan D B, Mills D and Vettier C 1988 *Phys. Rev. Lett.* **61**, 1241
[6] Hannon J P, Trammell G T, Blume M and Doon Gibbs 1988 *Phys. Rev. Lett.* **61**, 1245
[7] Templeton D H and Templeton L K 1994 *Phys. Rev.* B **49** 14850
[8] Cohen-Tannoudji C, Dupont-Roc J and Grynberg G 1988 in *Processus d'interaction entre électrons et atomes* chapter 2 (EDP Sciences and CNRS Editions, Les Ulis)
[9] Joly Y 2010 in *Magnetism and Synchrotron Radiation* New Trends Series: Springer Proceedings in Physics, **133** 77 Beaurepaire, E; Bulou H, Scheurer F and Kappler J-P (Eds.)
[10] Joly Y, Matteo S D, Bunău O 2012 *Eur. Phys. J. Special Topics* **208** 21-38
[11] Jackson, J. D., Classical Electrodynamics, 3rd ed. Wiley ed.
[12] Physicists and neutron crystallographists adopt this definition which is coherent with the time dependant Schrödinger equation $i\hbar\frac{d}{dt}\left|\psi(t)\right\rangle = H(t)\left|\psi(t)\right\rangle$. The other choice, $e^{i(\omega t-\mathbf{k}\cdot\mathbf{r})}$, is usual in the field of x-ray crystallography.
[13] Leininger P, Ilakovac I, Joly Y, Schierle E, Weschke E, Bunau O, Berger H, Pouget J P and Foury-Leylekian P 2011 *Phys. Rev. Lett.* **106**, 167203
[14] de Bergevin F and Brunel M 1981 *Acta Cryst.* A **37**, 314
[15] Blume M 1985 *J. Appl. Phys.* **57** 3615
[16] Blume M and Doon Gibbs 1988 *Phys. Rev.* B **37** 1779
[17] J.P. Hill and D.F. Mc Morrow 1996 *Acta Cryst.* A **52** 236-244
[18] Fabrizi F, Walker H C, Paolasini L, de Bergevin F, Boothroyd A T, Prabhakaran D and McMorrow D F 2011 *Phys. Rev. Lett.* **106** 239902

Resonant elastic X-ray scattering in life science, chemistry and materials science; recent developments

John R Helliwell

School of Chemistry, University of Manchester, M13 9PL, UK

john.helliwell@manchester.ac.uk

Abstract. The applications of anomalous scattering for locating metal atoms and discriminating between different elements has increased when optimised with synchrotron X-radiation. A distinct trend now is the introduction of in vacuum diffractometers to serve both the life and physical sciences. Examples of applications are given. Synchrotron source upgrades will allow nanosized X-ray beams to be more widely available. The new X-ray lasers suggest new capabilities too; first applications are described and new ones suggested involving harnessing heavy atom complexes for structure determination.

1. Introduction

Anomalous scattering results for locating metal atoms and discriminating between different elements has increased and found wide application in life science, chemistry and materials science from the optimisation via X-ray wavelength tuning with synchrotron radiation. The polarisation state of the synchrotron X-ray beam also allows dichroism effects to be explored, although the range of applications is less extensively developed as yet. The applications of resonant elastic X-ray scattering include the crystallography of microporous materials, superconductors and magnetic materials as well as in the biological chemistry of metalloproteins within life science.

With inorganic and materials chemistry, the number of wavelengths used is increasing and the ease of measuring diffraction data to usual chemical crystallography diffraction resolution standards has improved, even with longer wavelengths. Various instrumental and technical capabilities and case studies are described in [1]; a recent example reviewed there involved use of eleven X-ray wavelengths, each with diffraction data recorded to a usual, high, standard (*i.e.* at 0.84Å resolution) to study a zinc substituted gallophosphate, still a record it seems [2]. This ease of measurement of the resonant scattering signals along with absorption edge shifts indicates an expansion to measurement of multiple data sets to allow monitoring of redox changes, as a function for example of pH or temperature, which has already been done [2] and can expand. Whilst crystal structure determination in biological crystallography has been revolutionised by the synchrotron MAD (Multiple-wavelength Anomalous Dispersion) method [for a review see ref 3 and an example of MAD data measured in minutes see ref 4] MAD is not a requirement for chemical or materials crystallography as other phasing techniques are routine *i.e.* such as direct methods using single wavelength, non-resonant condition, diffraction data. Further applications to powders, solutions, thin films and amorphous materials indicate a buoyant contribution of Resonant Elastic X-ray Scattering (REXS) to chemistry and materials characterisation of atomic sites. Developments have included the use of high photon energies (upto 100 keV) as well as softer X-rays (2keV). The REXS 2011 article by the author with colleagues [1] reviewed these developments in detail. This article reviews developments and results from 2011 until now, and provides a short overview in the introduction. Other topics to highlight are:- applications of resonance diffraction to proteins and small molecules, including diffraction anomalous fine structure (DAFS) [5]; development of instrumentation and methods for MAD at various synchrotron radiation (SR) facilities [6]; and applications to small molecules and proteins, and an overview of complementary techniques [7].

A distinct trend now is the introduction of in vacuum diffractometers to serve both the life and physical sciences. Examples of applications are given. This trend, a natural one, in a way stems from the previous failure to enhance the anomalous dispersion due to difficulties reaching the longer wavelength absorption edges (see section 2); in chemical crystallography this still leaves the geometric limit of the diffraction resolution being never being able to be better than $\lambda/2$ even at full back scattering ($2\theta=180$ degrees) *i.e.* by Bragg's Law. A different technical frontier involves the new X-ray lasers and totally new capabilities beckon especially involving femtosecond time-resolved structural studies; first applications are described and new ones suggested involving harnessing heavy atom complexes to assist structure determination.

2. When do resonant scattering diffraction techniques fade or fail in capability? What is the remedy?

There are several situations where REXS techniques in crystallography are restricted or even fail. These are as follows:-

(i) Cases where lower atomic number metals are of interest (e.g. titanium which has a K-edge at 2.497 Å); even back scattering measurements of the X-ray reflections will 'only' yield 1.25 Å

Journal of Physics: Conference Series **519** (2014) 012002 doi:10.1088/1742-6596/519/1/012002

resolution); an alternative solution to obtaining site specific titanium occupancies has harnessed neutron diffraction of specifically substituted Ti isotopes in silicalite [8]

(ii) The case of even lower atomic number e.g. aluminium (7.948 Å) is interesting; ^{27}Al Nuclear Magnetic Resonance (NMR) studies of aluminosilicates and site-specific substitution are briefly reviewed in the context of REXS in [7];

(iii) Special apparatus for softer X-rays work, up to and including the phosphorus K edge at 5.784Å, was pioneered at the European Synchrotron Radiation Facility (ESRF) by H. Stuhrmann et al (reviewed in [9] and an in vacuum diffractometer, for macromolecular crystallography, is being built at Diamond Light Source by A Wagner et al, for example [see www.dls.ac.uk].

(iv) Radiation damage effects, as with all X-ray studies, can be a limitation or a challenge. A biological samples challenge is as exemplified in the photosynthesis protein 'PSII' studies [10]. In ASAXS (anomalous small angle X-ray scattering) studies of soft matter materials science radiation damage effects must be anticipated and minimised [11]. In our own previous research involving a nickel aluminophosphate we observed an X-ray radiation damage effect, admittedly after prolonged irradiation of 1 week at the National Synchrotron Light Source (NSLS) [12] and the impact of X-rays on disulphide bridges in proteins [13] is now well known.

3. Some future, general, trends

Dichroism effects are exploited already but will likely expand in the number of studies undertaken in the coming years, as appropriate software becomes routinely available [14]. Chemical changes induced by temperature and/or time-resolved studies should also increase the range and scope of these REXS applications [15]. In particular as the speed of measuring REXS improves, as more REXS capable instruments are available and as analysis is more routine then increasing numbers of samples will be studied per molecular system. Perhaps even sample to sample variations will be allowed for, although unlikely to be a major concern in general, but maybe of interest in specific cases. Protein powder diffraction harnessing dispersive differences is also under detailed investigation [16 and references therein]. A major initiative at 3rd generation SR sources, such as the ESRF Upgrade programme [see www.esrf.eu], emphasises nanosized X-ray beams, which when coupled with tunability will expand the REXS techniques to nanosized materials. Such nanosize X-ray beams can help look for local variations in micro-scans across a sample (although sample thickness will remain an averaging effect).

4. Softer X-rays REXS in life and physical sciences

A distinct trend now is the introduction of in vacuum diffractometers to serve both the life and physical sciences. In the physical sciences a relatively recent 4-circle, in-vacuum, diffractometer for resonant elastic soft x-ray scattering at the Canadian Light Source has been described [17]. It includes 9 in-vacuum motions driven by in-vacuum stepper motors and operates in ultra-high vacuum at a base pressure of 2 * 10^{-10} Torr. Cooling to a base temperature of 18 K is provided with a closed-cycle cryostat. The diffractometer includes a choice of 3 photon detectors: a photodiode, a channeltron, and a two dimensional sensitive channel plate detector. Along with variable slit and filter options, these detectors are suitable for studying a wide range of phenomena having both weak and strong diffraction signals. Example measurements of diffraction and reflectivity in Nd doped $La_{2-x}Sr_xCuO_4$and thin film (Ga,Mn)As have been described.

At the Diamond Light Source there is the in vacuum diffractometer for the physical sciences named 'RASOR' (See Figure 1) and ref [18].

A very recent review of resonant elastic soft x-ray scattering has been given by Fink et al [19]. This review emphasises that "resonant (elastic) soft x-ray scattering (RSXS) offers a unique element, site and valence specific probe to study spatial modulations of charge, spin and orbital degrees of freedom in solids on the nanoscopic length scale. The technique is not only used to investigate single-crystalline materials. This method also enables one to examine electronic ordering phenomena in thin films and to zoom into electronic properties emerging at buried interfaces in artificial heterostructures."

Equivalently, in the life sciences, the long-wavelength (1.5 to 4 Å) Diamond Light Source I23 project, led by Dr Armin Wagner, will be a user facility to enhance and optimise the anomalous signals from low atomic number elements. These include include sulphur in proteins and/or phosphorous in RNA/DNA (ribonucleic acid/deoxyribonucleic acid) crystals, needed where protein labelling to introduce anomalous scatterers is not feasible, such as involving Se methionine via molecular biology gene expression or heavy atom chemical derivatisation. In addition, for the Diamond Light Source I23 the beamline's wavelength range will provide access to the M-edges of

Journal of Physics: Conference Series **519** (2014) 012002 doi:10.1088/1742-6596/519/1/012002

elements, *e.g.* of uranium, with huge anomalous signals. A large semi-cylindrical detector and a multi-axis goniometer will be available for crystal alignment and orientation. X-ray tomography to obtain the crystal sample shape and volume for an analytical absorption correction will be possible.

Figure 1 The in vacuum 'RASOR' diffractometer at the Diamond Light Source [18]. Courtesy of the website of the Diamond Light Source and to whom we are grateful.

The figure 2 below illustrates with chlorine as the element the long wavelength, for crystallography, of the absorption edge and the f ' and f " variations.

We have recently published on another practical situation where the raw X-ray diffraction data proved valuable and are now briefly reviewed. This is the example [20)] of locating a low Z anomalous scatterer at fractional occupancy in a pharmaceutical chemistry crystallography study. Tanley et al 2014 [20] describe a partial chemical conversion of carboplatin to cisplatin under a high (10%) NaCl condition. This meant that they had the crystallographic challenge to see a partially occupied chlorine at the wavelength used for X-ray diffraction data collection of 1.54 Å (*i.e.* with a quite small f " of 0.7 electrons for a fully occupied Cl). This study showed the importance of open archiving of the diffraction data images and which allowed a wider comparison of software results from them than the original study (for details see [21].

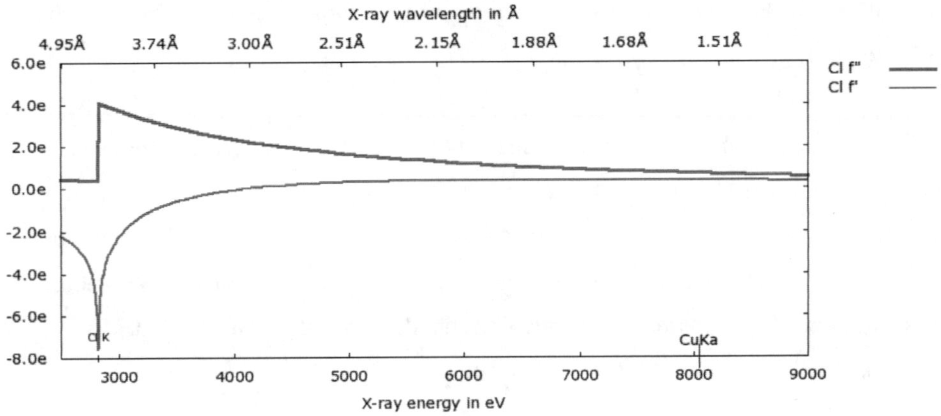

Figure 2 The variation of the REXS f " and f ' values with X-ray wavelength and photon energy for chlorine (theoretical values) as an example low Z element of specific interest in the life sciences for *e.g.* structural studies of protein salt (NaCl) interactions or pharmaceutical compounds like cisplatin (see Figure 3). [Courtesy of the website http://skuld.bmsc.washington.edu/scatter/ and Prof Ethan Merritt, to whom we are grateful.]

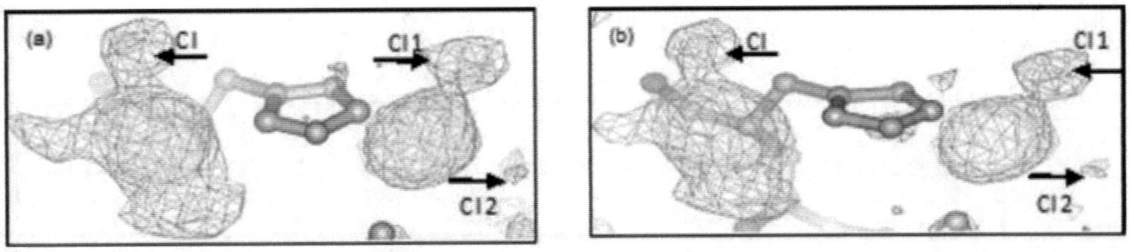

Figure 3. The two binding sites on the His-15 residue of hen egg white lysozyme (HEWL). The Fo-Fc omit electron density maps are in green and the anomalous difference electron density maps are shown in orange at 3σ cut off; the diffraction data for the Protein Data Bank code 4dd7 study were processed by different software namely (a) XDS [22] and (b) EVAL [23]. From [20] with the permission of the IUCr (http://journals.iucr.org/).

Low Z atomic number elements are especially challenging for protein X-ray crystallography. The classic case is the problem of trying to identify a sodium ion versus a bound water. Such a case in the structural chemistry details can be resolved by the putative sodium ligand distances (~2.2 A) and the expected octahedral environment versus a hydrogen bonding interaction for a bound water ((~2.8 A) provided adequate structural precision is available, which of course is not always the case. Besides software improvements for diffraction data image processing of course experimental developments continue apace. Thus the expansion into use of longer X-ray wavelengths is a real help too in such cases *e.g.* see [24] for a short summary.

Finally, W A Hendrickson and colleagues [25] have recently devised robust procedures for enhancing the signal-to-noise ratio in the anomalous scattering from native biological structures by combining diffraction data measured from multiple crystals at lower-than-usual X-ray energy, both for structure determination and element identification. At the upcoming new National Synchrotron Light Source 'NSLS2' there is a proposed beamline for longer wavelengths' data collection (W A Hendrickson presented at European Crystallography Meeting (ECM) 28).

5. Applications with the X-ray laser

This seems challenging territory for X-ray diffraction in general and REXS in particular. We must ask firstly, most basically, how might the X-ray atomic scattering factor be affected with time in the X-ray laser beam? We take guidance from [26] and Figure 4, which shows the population dynamics for several selected charge states of iron, as an example, during an X-ray pulse; how the X-ray laser pulse affects the scattering factors in a sample is a matter of future practical research in structure determination.

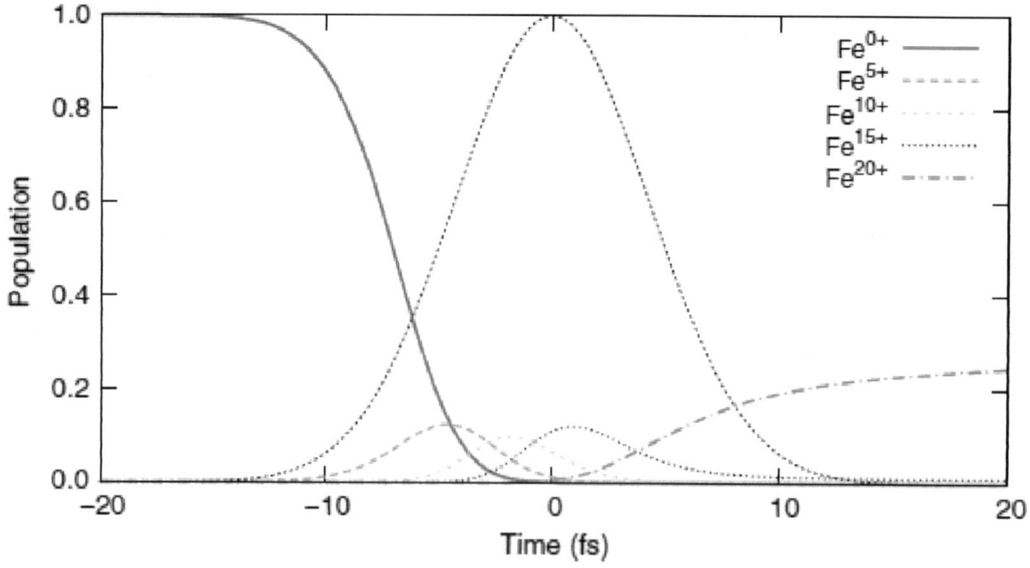

Figure 4 Population dynamics for several selected charge states of Fe during an X-ray pulse. The thin dotted line indicates the X-ray pulse envelope assumed Gaussian. Reprinted figure 1 with permission from ref 26 namely S.-K. Son, H. N. Chapman, and R. Santra, Phys. Rev. Lett. 107, 218102, 2011. Copyright (2011) by the American Physical Society and with permission from the authors to whom we are grateful.

The X-ray laser has been used to study the visibility of the anomalous dispersion signal from protein disulphides [27a]. This is a particularly telling example for life scientists and for which specific X-ray damage has been carefully investigated over the last couple of decades. Barends et al [27a] successfully saw the anomalous signal from sulphur atoms in protein crystallographic data measured with an X-ray free-electron laser; see figure 5. Barends et al commented that "the results presented clearly demonstrate that despite the obvious challenges associated with the analysis and evaluation of diffraction data collected with (X-ray) free-electron laser sources, anomalous (difference f ") data can be obtained by serial femtosecond (X-ray) crystallography." In a most recent development Barends et al 2014 [27b] study the 'Single-wavelength Anomalous Dispersion (SAD) phasing' effectiveness for a gadolinium derivatised lysozyme, using X-ray laser diffraction data from micron sized crystals and who also make a comparison with a 'zero X-ray dose' home laboratory X-ray source single 'normal sized' crystal dataset. The gadolinium anomalous signal seen within the X-ray laser diffraction data is remarked upon by Barends et al 2014 as being considerably lower than expected. The authors attribute this to an insufficient number of X-ray laser diffraction images being measured, due to a limited beamtime, rather than an effect on the anomalous scattering signature itself. Further X-ray laser experiments are obviously needed to investigate this aspect further.

In [24] the suggestion was made for use of marker heavy atoms to assist with structure determination with X-ray Free Electron Laser (XFEL) diffraction data. "There may also remain interest in the use of XFEL wavelengths at the xenon L edges *e.g.* to control the value of xenon f ' values for protein structure determination where differences at two wavelengths in single molecule diffraction data could reveal the location of xenon atoms. A similar two wavelength, delta f ', approach would work for iodine or sulphur to locate their positions in a protein single molecule imaging experiment as well. Such stepwise structural approaches may prove valuable in challenging single molecule protein structure determinations via the XFEL route." Whilst the focus today is the use of protein microcrystals, and then nanocrystals, at the X-ray laser the ultimate target sample size of single molecules for diffraction [28] is of keen interest. An intermediate step beyond nanocrystals, before the single molecule sample, would be a nanocluster of molecules. A natural extension of marker atoms in such structural studies with the X-ray laser, presented by J R Helliwell at European Crystallography Meeting (ECM) 28, building on protein two wavelength powder MAD diffraction developments [16], would be to use heavy metal complexes such as the platinum hexahalides, most notably the platinum hexa-iodide and the platinum hexa-bromide, and or tantalum bromide (Ta_6Br_{12}). These would provide recognisable shapes in electron density maps, respectively octahedral for the platinum hexahalides and roughly spherical for tantalum bromide, and also enhance the overall X-ray scattering strength of a single molecule or nanocluster type of sample size challenge. The particular chemistry for their binding of these metal complexes, namely a complementary shape and or electronic charge to a binding site on the protein, would restrict their applicability of course. Another possibility would be iodine triangles [29] albeit less electron rich but another recognisable shape that would be helpful where there was a noisy electron density map to be interpreted; these offer a further range of chemistry binding site types [29].

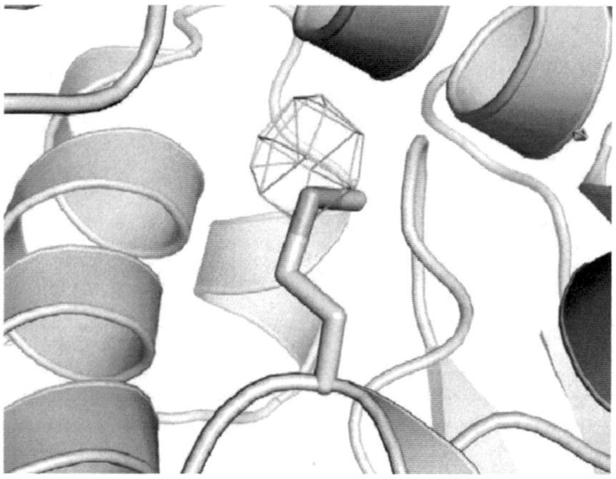

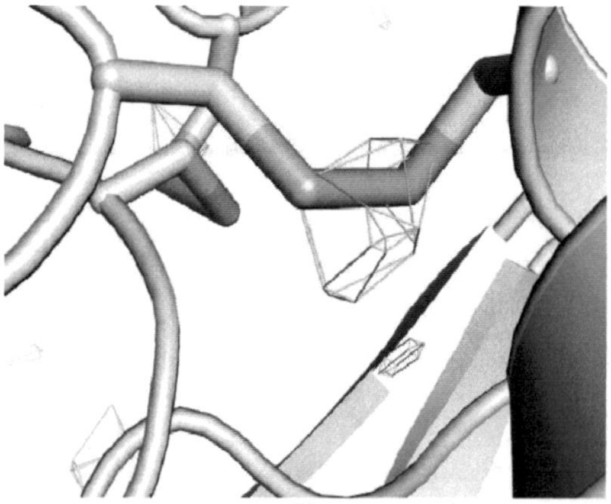

Figure 5 Anomalous difference Fourier maps (yellow) based on data measured at the LCLS for (a) Met105 contoured at 3σ and (b) the Cys76-Cys94 disulfide bond contoured at 2.5σ. From [27a] and refs therein with permission of the IUCr (http://journals.iucr.org/) and the authors to whom we are grateful.

6. Concluding remarks and future prospects

The field of application of anomalous scattering in structural chemistry and biology is now truly vast. The evolution of capability is breath taking and shows up the pace of modern day research.

Overall, to conclude, the necessary technical advances have been made to harness REXS X-ray anomalous scattering at synchrotron radiation sources these last decades, which have now brought reliable answers to a vast number of life science, bioinorganic, inorganic and materials chemistry research challenges.

The future indicates an expansion of the field as the measurements are more rapid and there are increased numbers of such tunable SR beamlines available. Beamlines providing high and low photon energies (in a range of ~90 to 2 keV) indicate wide chemical element capability. The SR source upgrades towards providing nanosized tunable X-ray beams, such as at ESRF (www.esrf.ac.eu), opens new possibilities of applying REXS techniques increasingly for nanoscience. Also the XFELs for REXS applications in structural chemistry and biology, even perhaps with single molecules and or nanoclusters, are an interesting prospect if the wavelength setting of the X-ray laser can be controlled and the inner shell photoelectron ejection [30] proves not to be limiting.

12. Acknowledgments

J R Helliwell thanks all his past and current collaborators over the years in these fields. Numerous funding agency and SR beamtime awards have been received including from the Science and Engineering Research Council, the Swedish Research Council, the Biotechnology and Biological Sciences Research Council, the Engineering and Physical Sciences Research Council and the Council for the Central Laboratories of the Research Councils, which are the UK government research agencies, and the Synchrotron Radiation Source (SRS) Daresbury, the NSLS, the Sincrotrone Trieste 'ELETTRA', the ESRF and Diamond Light Source and to whom JRH is especially grateful. JRH's anomalous scattering work has evolved over a period of 30 years and so he is grateful to the

Universities of Oxford, Keele, York and Manchester and to the Daresbury Laboratory for providing stimulating environments in which to study and to do experiments. JRH also thanks The Royal Society for a Research Exchange Agreement funding travel and subsistence with Prof V Kaucic and Dr N Zabukovec Logar at the National Institute of Chemistry in Ljubljana, Slovenia.

References

[1] Helliwell J R, Helliwell M, Kaucic V, and Logar N Z 2012 Eur. Phys. J. Special Topics 208, 245-257
[2] M Helliwell, J. R. Helliwell, V. Kaucic, N. Zabukovec Logar, S. J. Teat, J. E. Warren and E. J. Dodson 2010 Acta Cryst B66, 345-357
[3] Walsh M A, Evans G, Sanishvili R, Dementieva I and Joachimiak A 1999 Acta Cryst. D55, 1726-1732
[4] Walsh M A, Dementieva I, Evans G, Sanishvili R and Joachimiak A 1999 Acta Cryst. D55, 1168-1173
[5] Favre-Nicoli V, Proietti M G, Leclere C, Katcho N A, Richard M-I and Renevier H Eur. Phys. J. Special Topics 208,189–216(2012)
[6] Cassetta A, Deacon AM, Ealick S E, Helliwell J R and Thompson AW 1999 J. Synchrotron Rad. 6, 822–833
[7] Cianci, M, Helliwell, J R, Helliwell, M, Kaucic, V, Logar, N Z, Mali, G and Tusar, N N 2005 Crystallography Reviews,11:4, 245 — 335
[8] Henry P F, Weller M T and Wilson C C 2003 J. Appl. Cryst. 36, 1361
[9] Djinovic Carugo K, Helliwell J R, Stuhrmann H and Weiss M S 2005 J. Synchrotron Rad. 12, 410-419
[10] Umena Y, Kawakami K, Shen J-R, Kamiya N 2011 Nature 473, 55-60
[11] Sztucki M, Di Cola E, Narayanan T 2010 J. Appl. Cryst. 43, 1479-1487
[12] Helliwell M, Gallois B, Kariuki B, Kaucic V and Helliwell J R 1993 Acta Cryst. B49, 420–428
[13] Helliwell J R 1988 J. Crystal Growth 90, 259-272
[14] Schiltz M and Bricogne G 2010 Acta Cryst. D66, 447-457
[15] Wu G, Zhang Y, Ribaud L, Coppens P, Wilson C, Iversen B B and Larsen FK 1998 Inorg. Chem. 37, 6078-6083
[16] Helliwell J R, Bell A M T, Bryant P, Fisher S J, Habash J, Helliwell M, Margiolaki I, Kaenket S, Watier Y, Wright J and Yalamanchilli S 2010 Z Krist 225 570-575
[17] Hawthorn, D G et al 2011 The Review of scientific instruments Volume: 82 Issue: 7 Pages: 073104 DOI: 10.1063/1.3607438
[18] http://bragg.phycmx.dur.ac.uk/wiki/projects/rasor
[19] Fink J, Schierle E, Weschke E and Geck, J 2013 Reports on Progress in Physics 76 (5) Pages: 056502
[20] Tanley S WM, Diederichs K, Kroon-Batenburg L M J, Schreurs A M M and Helliwell JR 2013 Journal of Synchrotron Radiation 20, 880–883.
[21] Tanley S W M, Schreurs A M M, Helliwell J R and Kroon-Batenburg L M J 2013 J. Appl. Cryst. 46, 108-119
[22] Kabsch, W 2010 Acta Cryst D 66:125-132
[23] Schreurs, A M M, Xian, X & Kroon-Batenburg, L M J. 2010 J.Appl. Cryst. 43, 70–82
[24] Helliwell J R 2004 J. Synchrotron Rad. 11, 1-3

[25] Liu, Q, Dahmane, T, Zhang, Z, Assur, Z, Brasch, J, Shapiro, L, Mancia, F, Hendrickson, W A 2012 Science 336, 1033-1037.
[26] Son, S K, Chapman, H N and Santra, R 2011 Phys Rev Lett 107, 218102
[27] a. Barends T R M et al 2013 Acta Cryst. D69, 838-842; b. Barends T R M, Foucar L, Botha S, Doak R B, Shoeman R L, Nass K, Koglin J E, Williams G J, Boutet S, Messerschmidt M & Schlichting I Nature 505, 244–247
[28] Neutze R, Wouts R, van der Spoel D, Weckert E and Hajdu J 2000 Nature 406 752-757
[29] Beck, T, Krasauskas, A, Gruene, T and Sheldrick, GM 2008 Acta Cryst 64, 1179-1182.
[30] Hau-Riege, S P 2007 *Phys. Rev. A* 76, 042511

Journal of Physics: Conference Series **519** (2014) 012003 doi:10.1088/1742-6596/519/1/012003

An X-ray study of the Dzyaloshinskii-Moriya interaction in the weak ferromagnet $FeBO_3$

V E Dmitrienko[1], E N Ovchinnikova[2], S P Collins[3], G Nisbet[3] and G Beutier[4]

[1] Institute of Crystallography,Moscow, 119333, Russia
[2] Physical Department, Moscow State University, 119991, Russia
[3] Diamond Light Source Ltd, Diamond House, Harwell Science and Innovation Campus, Didcot, Oxfordshire, OX11 0DE, UK
[4] SIMaP, UMR 5266 CNRS, Grenoble-INP UJF, France

Abstract. We report on the axis, magnitude and direction of the Dzyaloshinskii-Moriya (DM) interaction in the weak ferromagnet $FeBO_3$. The latter relies on the determination of the phase of the magnetic x-ray scattering amplitude. We outline a new technique based on interference with forbidden quadrupole resonant scattering to obtain this phase information.

1. Introduction

Weak ferromagnetism in materials such as $\alpha-Fe_2O_3$ was cited by Moria [1] as being '*a controversial problem for a decade*' back in 1960, and played a crucial role in the development of the theory of antisymmetric exchange (Dzyaloshinskii-Moriya) interactions. In recent years interest has shifted towards other manifestation of the DM interaction, such as the magnetoelectric effect [2] and skyrmion states [3], where reliable theoretical predictions have been lacking. Weak ferromagnets provide an excellent test-bed for theories due to the fact that the twisting of the atomic moments is purely local and leads to a small and manageable magnetic unit cell, in contrast to the long-period modulations observed in globally chiral crystals. Unfortunately, while the local nature of the twisted moments is highly beneficial for theoretical development, it leads to some severe experimental challenges. Most notably, it is necessary to extract the *phase* of the magnetic scattering, rather than the usual intensity.

In this paper, we present a study of the weak ferromagnet $FeBO_3$. We discuss the direction (axis) of the DM interaction by considering its symmetry, and the magnitude of the interaction via the canting angle. Most significantly, we show the first results from a novel interference technique proposed by some of us [4] to determine the phase of the magnetic x-ray scattering signal, and the sign of the DM interaction in $FeBO_3$.

2. Dzyaloshinskii-Moriya Interaction and Symmetry

The most general form of the bilinear coupling energy between two spins, $\mathbf{S}^1$ and $\mathbf{S}^2$ (S_i^1 and S_j^2 in Cartesian tensor component form) can be written as the contraction of a second-rank tensor M_{ij} with the spins to form a phenomenological classical spin Hamiltonian,

$$\Delta E = S_i^1\, S_j^2\, M_{ij} \tag{1}$$

Journal of Physics: Conference Series **519** (2014) 012003 doi:10.1088/1742-6596/519/1/012003

Here, the nine components M_{ij} can be expressed as a scalar (one component), a symmetric traceless tensor (five components) and an antisymmetric tensor (three components):

$$\Delta E = J S_i^1\, S_j^2\, \delta_{ij} + S_i^1\, S_j^2\, M_{ij}^S + S_i^1\, S_j^2\, M_{ij}^A \tag{2}$$

Of these three terms, the isotropic part always dominates and the symmetric anisotropic term is small enough that it can typically be neglected. The third term (antisymmetric part) is the *Dzyaloshinskii – Moriya* interaction, which is similarly small but plays a crucial role in the canting of the spins. The isotropic and antisymmetric parts can be written equivalently in terms of scalar and vector products of the spins, respectively:

$$\Delta E = J\mathbf{S}^1 \cdot \mathbf{S}^2 + \mathbf{D} \cdot [\mathbf{S}^1 \times \mathbf{S}^2] \tag{3}$$

Table 1. The Moriya Rules [1] that govern the DM vector, **D**, between two spins at points A and B with a mid-point at C, and a comment on whether these are consistent with **D** transforming as a polar or axial vector

Moriya Rule	Polar?	Axial?
1. When a center of inversion is located at C, $\mathbf{D} = \mathbf{0}$	yes	no
2. When a mirror plane perpendicular to AB passes through C, $\mathbf{D} \parallel$ mirror plane or $\mathbf{D} \perp AB$	yes	no
3. When there is a mirror plane including A and B, $\mathbf{D} \perp$ mirror plane	no	yes
4. When a two-fold rotation axis perpendicular to AB passes through C, $\mathbf{D} \perp$ two-fold axis	no	no
5. When there is an n-fold axis ($n \geq 2$) along AB, $\mathbf{D} \parallel AB$	yes	yes

If we consider the strength of the coupling between spins to be a physical property of the crystal then **D** might be expected to possess the symmetry of the environment in which it acts, specifically, the site-symmetry of the point mid-way between the coupled spins. As **D** is clearly a vector, one might then ask whether it transforms as a polar or axial vector. Indeed, Moriya studied the symmetry of the antisymmetric exchange interaction in a seminal paper [1] that introduced five celebrated 'Moriya rules'. We reproduce those rules in Table 1 and ask, for each, whether the rule is compatible with a polar or axial vector. Interesting, we see that no clear pattern emerges: it appears that **D** does not transform as a vector at all!

The difficulty arises from the fact that our Hamiltonian neglects the vector that links the positions of the two spins - an omission that is critical due to the fact that the positions of the spins is reversed by some symmetry operations. Dmitrienko et al. [4] argued that to describe the DM coupling with correct transformation properties, one must introduce a third-rank tensor, which couples to the two spins as well as the vector linking them:

$$\Delta E = T_{ijk}\, S_i^1\, S_j^2\, \hat{r}_k^{12} \tag{4}$$

from which the DM 'vector' can be obtain from

$$D_n = \frac{1}{2}\varepsilon_{ijn}\, T_{ijk}\, \hat{r}_k^{12} \tag{5}$$

where ε_{ijn} is the permutation tensor. As an indication that such a procedure is not unreasonable, one can apply the symmetries highlighted by the Moriya rules. In all five cases the representation developed by Dmitrienko et al. reproduce the results of Moriya. Moreover, from Eqn. 5 one can obtain the sixth Rule omitted by Moriya: vector **D** is directed along $\hat{\mathbf{r}}^{12}$ when there is

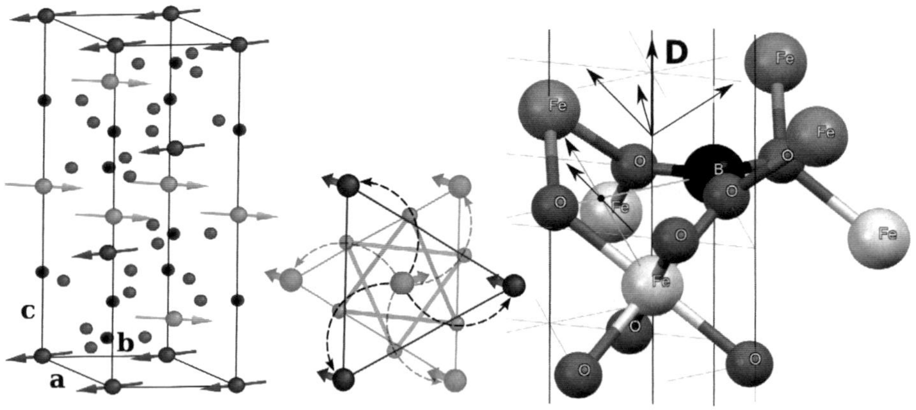

Figure 1. Magnetism in $FeBO_3$ (spacegroup $R\bar{3}c$). Left: a magnetic (hexagonal) unit cell, showing oxygen atoms (red), boron atoms (black), and two symmetry-related magnetic iron sublattices (blue and grey) with moments tilted in the ab plane between the two. Right: DM vectors. The vector linking two neighbouring irons is shown in grey, with a two-fold axis (green) passing through its mid-point (black dot). The fourth Moriya rule dictates that the DM vector (thin black arrow) must be perpendicular to the two-fold axis. The resultant DM vector linking the two sublattices (thick black arrow) is the sum of individual vectors, which are related by three-fold rotation along c (thin back lines), and therefore points along the three-fold axis. Centre: The twisting of oxygen atoms and magnetic spins, going up (green oxygens, dark blue iron atoms) and down (pink oxygens, light blue iron atoms) from the central iron atom, looking into the c axis.

an n-fold ($n \geq 3$) axis perpendicular to $\hat{\mathbf{r}}^{12}$ (i.e. vector $\mathbf{D}$ is normal to the n-fold axis - a somewhat counterintuitive result!). For example, if two spins interact through a chiral liquid then, according to spherical symmetry of liquid, $T_{ijk} \propto \varepsilon_{ijk}$ and $\mathbf{D} \parallel \hat{\mathbf{r}}^{12}$.

Having established that the transformation properties of the third-rank DM *tensor* reproduce the Moriya rules, one can carry out the necessary symmetrization for the site symmetry of all mid-points between neighbouring magnetic ions in order to establish (a) whether or not the DM interaction is allowed and (b) any symmetry constraints on its orientation. In the case of $FeBO_3$, the nearest iron atoms (one from each sublattice) have at their mid-point a non-centrosymmetric site, allowing in principle the possibility of the DM interaction. Moreover, a two-fold axis, perpendicular to the two iron atoms passes through the midpoint. Moriya Rule No. 4 states that $\mathbf{D}$ must therefore act perpendicular to the two-fold axis (Fig. 1). Finally, summing over all symmetry-equivalent $\mathbf{D}$ vectors that act between the two iron sublattices, three fold rotation symmetry about the c axis ensures that the resultant $\mathbf{D}$ vector is parallel to c. A DM vector along c is consistent with the proposed magnetic structure: from Fig. 1 we see that the vector product of a spin from each sublattice is directed along c.

3. The Magnitude of the Dzyaloshinskii-Moriya Interaction

The DM interaction is weak compared to the isotropic exchange, meaning that the magnetic structure of $FeBO_3$ is essentially antiferromagnetic, with sublattices almost antiparallel and a small twist leading to *weak* ferromagnetism. The classical spin Hamiltonian in Eqn. 3 can then be re-written in terms of the canting angle (ϕ) between sublattice,

$$\Delta E \propto -J\cos(2\phi) + D\sin(2\phi). \quad (6)$$

As this state must correspond to an energy minimum we can set its derivative to zero to obtain

$$\frac{D}{J} = -\tan(2\phi) \simeq -2\phi \quad (7)$$

The value of the isotropic exchange, obtained by Mössbauer spectroscopy, and the magnitude of the canting angle determined by magnetization measurements, are J=7.5 meV and $|\phi|$= 0.016 rad. respectively [5]. Inserting these values into Eqn. 7 gives the magnitude of the DM vector to be $|D|$=0.24 meV.

4. The Sign of the Dzyaloshinskii-Moriya Interaction

At this point, we already know *almost* everything about the DM vector: we know its magnitude, and we know that it lies parallel to the crystal c axis. What we *don't* know is whether the DM vector linking adjacent sublattices points up or down. That is, we don't know the sign of the DM-vector, i.e. whether the magnetic twist follows the twist of the oxygen atoms between magnetic ions or opposes it.

The standard techniques for determining the arrangement of atomic spins in a crystal are x-ray and neutron diffraction. Assuming that orbital polarization is extremely small, the structure factor for the non-resonant magnetic $(hkl) = (0, 0, 6n + 3)$ reflections can be written [6] as,

$$\mathbf{F}_{\text{spin}}(\mathbf{Q}) = 2Sf_S(\mathbf{Q})\frac{\mathbf{H} \times \mathbf{D}}{|\mathbf{H}||\mathbf{D}|} \quad (8)$$

where S is the total spin of the iron atom, $f_S(\mathbf{Q})$ is the spin form factor, and $\mathbf{H}$ is the direction of a weak external field, strong enough to rotate the ferromagnetic component into the field direction. This expression shows that the magnetic structure factor, which is determined by the antiferromagnetism, is perpendicular to $\mathbf{H}$ as one would expect from Fig. 1. The ability to rotate the antiferromagnetic structure via an external field is extremely attractive as it allows structure factor vector directions to be determined without physically rotating the crystal in an 'azimuthal' scan. However, since the scattered intensity $I \propto |\mathbf{B} \cdot \mathbf{F}_{\text{spin}}(\mathbf{Q})|^2$, there is an obvious problem: the sign of $\mathbf{D}$ reverses the phase of the magnetic scattering but does not affect its intensity. (Here, vector $\mathbf{B}$ depends on the polarization and wave vectors of incident and scattered beams; for the $\sigma - \pi$ channel used in our experiments, $\mathbf{B}_{\sigma\pi} = \mathbf{k}(1 - \mathbf{k} \cdot \mathbf{k}')$).

In order to determine the phase of the magnetic scattering and hence the sign of the DM interaction, it is necessary to carry out some kind of interference measurement. It was pointed out by Dmitrienko et al. [4] that in $FeBO_3$, and indeed a surprisingly wide range of other weak ferromagnets, there is a fortuitous overlap between magnetic diffraction and pure quadrupole resonant forbidden x-ray scattering, of the type first reported by Finkelstein et al. [7]. The latter is only slightly stronger than the magnetic scattering and occurs at almost a single sharp resonance in the iron K pre-edge region. The scattered intensity is now given by $I \propto |\mathbf{B} \cdot \mathbf{F}_{\text{spin}}(\mathbf{Q}) + F(E2E2)|^2$ where $F(E2E2)$ is the pure quadrupole scattering amplitude. Moreover, despite it's apparently exotic nature, $F(E2E2)$ is very well understood and accurately modeled by theoretical codes such as FDMNES [8] and thus provides a reliable 'reference wave' for an interference experiment.

Experiments were carried out at XMaS (BM-28) at the ESRF, and I16 at Diamond Light Source. In both cases the sample was attached to a cryofurnace, mounted on a large six-circle diffractometer. At I16, measurements were made of the 003 forbidden reflection close to resonance, revealing a single sharp resonant peak that is aligned with the pre-edge feature of the fluorescence spectrum (Fig. 2, left), as is typical for pure quadrupole resonant reflections. Measuring far from resonance allowed observation of the much weaker pure magnetic reflection, also at the 003 position, originating from the antiferromagnetic modulation of the iron moments.

Journal of Physics: Conference Series **519** (2014) 012003 doi:10.1088/1742-6596/519/1/012003

The magnetic motif was rotated by a small (0.011 T) external field to produce a very high quality 'azimuthal' type scan, with a shape that is consistent with pure spin scattering (Fig. 2, right). It is worth noting that, the high quality of the data obtained via this technique is due mainly to the fact that the sample azimuthal angle is fixed and so no multiple scattering conditions are excited during the measurement.

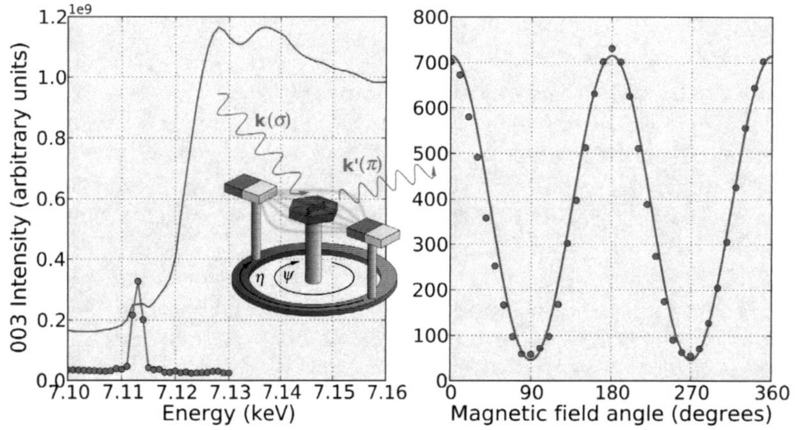

Figure 2. Left: Fluorescence (green line), and the 003 resonant diffraction spectra (red lines and circles) showing a single sharp peak. Right: The magnetic field orientation dependence of the 003 magnetic reflection far below resonance at 5.1 keV (red circles) and the calculated field dependence, assuming spin-only scattering (green). Insert: A schematic of the experiment showing the rotating magnet assembly and $FeBO_3$ spins.

In order to observe interference between the two scattering processes, XMaS measurements were carried out at the 009 reflection position, where the amplitudes were more similar in magnitude than the 003. A polarization analyzer stage was employed to select only the $\sigma - \pi$ channel, again to maximize interference, since the resonant amplitude is entirely in $\sigma - \pi$, while the magnetic scattering is mixed. The origin of the rotations shown in Fig. 2 are such that $\psi = 0$ and η=0 correspond to the crystallographic (100) axis and field direction lying parallel to $\mathbf{k} + \mathbf{k}'$.

As the major result of the present study was a sign, a great deal of care was taken to ensure consistency of sign conventions, the coordinate systems of measurements and simulations, and the direction of the applied magnetic field. The latter was determined with respect to the Terrestrial field.

The results shown in Fig. 3 correspond to interference between the resonant and magnetic amplitude, taken with two opposite field directions to cause a reversal of the magnetic amplitude between scans. One can see a small but clear jump in the energy of the resonant scattering peak as the magnetic field is rotated by 180°, causing the sign of the magnetic scattering, which is given by a projection of the magnetic structure factor in Eqn. 8, to reverse. Such a jump can be understood in terms of constructive/destructive interference between resonant and magnetic scattering on the low/high energy side of the resonance. The direction of the jump gives the phase of the magnetic scattering. The phase of the magnetic scattering gives the sign of the DM interaction. We find that the the magnetic twist follows the twist in the intermediate oxygen atoms in the planes between the iron planes, as depicted in Fig. 1.

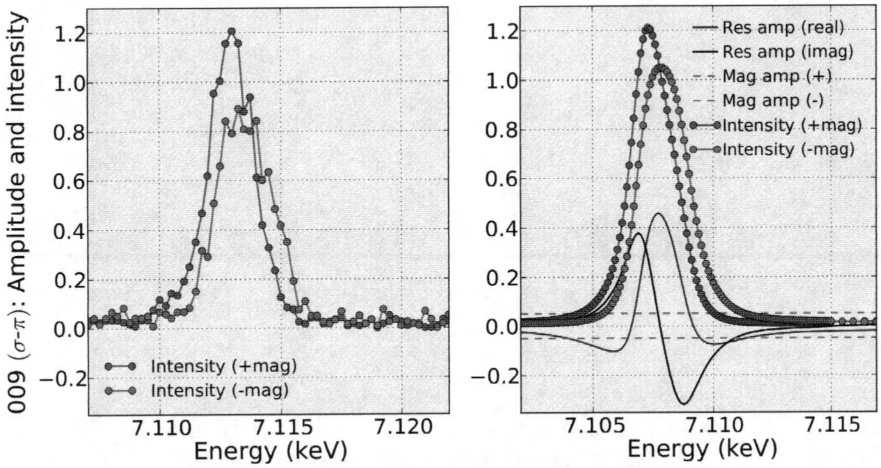

Figure 3. The measured (left) and simulated (right) 009 diffraction intensity close to resonance with the magnetic field applied in two opposite directions to reveal a 'jump' in the resonance energy. Simulations based on the FDMNES program [8] show how the direction of the jump depends on the phase of the magnetic scattering, which in turn depends on the sign of the DM vector. The jump is caused by interference between the antisymmetric imaginary (black line) part of the resonant scattering amplitude - the result of a superposition of two very close resonances of almost equal amplitude but opposite sign - and the purely imaginary magnetic scattering (dashed lines).

5. Conclusions

We have studied the DM interaction in $FeBO_3$, using a variety of techniques to determine its axis, strength and sign. The latter represents the first use of a novel interference technique. These results provide a very sensitive test for the development of predictive theories that can be applied to a range of phenomena including weak ferromagnetism, magnetoelectric materials and skyrmion states.

References

[1] Moriya T (1960) *Phys. Rev.* **120** 91
[2] Cheong S.-W. and Mostovoy M (2007) *Nature Mater.* **6** 13
[3] Rößler U K Bogdanov A N and Pfleiderer C (2006) *Nature* **442** 797
[4] Dmitrienko V E, Ovchinnikova E N, Kokubun J and Ishida K (2010) *JETP Lett.* **92** 383
[5] Eibschütz M and Lines M E (1973) *Phys. Rev.* B **7** 4907
[6] Dmitrienko V E, Ovchinnikova E N, Collins S P, Nisbet G, Beutier G, Kvashnin Y O, Mazurenko V V, Lichtenstein A I and Katsnelson M I (2013) *Nature Physics (at press)*
[7] Finkelstein K D, Shen Q and Shastri S (1992) *Phys. Rev. Lett.* **69** 1612
[8] Joly Y (2001) *Phys. Rev.* B **63** 125120

Magnetic phase diagram and ordered ground state of $GdMn_2O_5$ multiferroic studied by x-ray magnetic scattering

C. Vecchini[1,2], A. Bombardi[1], L. C. Chapon[3,4], N. Lee[5], P. G. Radaelli[6] and S-W. Cheong[5]

[1]Diamond Light Source Ltd, Harwell Science and Innovation Campus, Didcot, Oxfordshire, OX11 0DE, UK
[2] National Physical Laboratory, Hampton Road, Teddington, TW11 0LW, UK
[3] Institut Laue-Langevin, BP 156X, 38042, Grenoble, France.
[4] ISIS Facility, Rutherford Appleton Laboratory - STFC, OX11 0QX, Chilton, Didcot United Kingdom
[5] Rutgers Center for Emergent Materials and Department of Physics & Astronomy, Rutgers University, Piscataway, New Jersey 08854, USA.
[6] Department of Physics, University of Oxford, Parks Road, Oxford OX1 3PU, United Kingdom

Abstract. The magnetic structure of multiferroic $GdMn_2O_5$ is studied by x-ray magnetic scattering, both off resonance and in resonant conditions at the Gd L_{III} edge. Temperature dependence of magnetic Bragg reflections shows an initial incommensurate (ICM) ordering appearing at T$\sim$40K with magnetic propagation vector $\mathbf{k}_{ICM}$ $\sim$(0.49 0 0.18) which condenses at T$\sim$33K in a commensurate (CM) wave vector $\mathbf{k}_{CM}$=($\frac{1}{2}$ 0 0). In the CM phase Gd^{3+} ions appear to order spontaneously at the same temperature as Mn ions. Azimuthal scans allowed for the precise determination, from x-ray magnetic scattering only, of the orientation and magnitude of the Gd^{3+} magnetic moments. The Gd^{3+} magnetic isotropy allows for the Gd ions to mimic the Mn^{4+} magnetic sublattice via Gd-Mn-Gd superexchange interactions.

1. Introduction

Spin-driven ferroelectrics have attracted much interest due to the possibility of controlling their electric (magnetic) polarization by applying a magnetic (electric) field[1, 2]. Over the last few years, the experimental and theoretical studies of very diverse materials allowed to identify several magnetoelectric coupling mechanisms at play, driven for example by symmetric and antisymmetric magnetic exchange terms or single-ion effects. In the so-called symmetric exchange-striction model, polar displacements are due to a structural relaxation induced by competing Heisenberg terms ($\mathbf{J}_{ij}\mathbf{S}_i\cdot\ \mathbf{S}_j$) in the magnetic Hamiltonian. This mechanism can lead to large electrical polarization (**P**) in type-II multiferroics (compared with that of proper ferroelectrics such as $BaTiO_3$ is still minuscule) and has been proposed in the case of the RMn_2O_5 compounds (R=rare-earth, Y,Bi)[3], $Ca_3CoMnO6$[4] and the manganite E-phase[5]. Of importance here is that this model maximizes **P** for commensurate magnetic order and collinear spins.

The RMn_2O_5 compounds have been extensively studied since the seminal work on the Tb analogue showing that **P** can be reversed by an applied magnetic field[6]. For this class of materials, it is now accepted, after much debate, that the leading mechanism is related to symmetric-exchange striction induced by magnetic ordering of the Mn. A universal feature of the magnetoelectric phase diagram (for R=lanthanide) is the complex sequence of phase transitions on cooling below $\sim$ 40K[7]: first a paraelectric phase with incommensurate magnetic order, followed by a ferroelectric phase (b polar-axis) with commensurate magnetic order (wave-vector $\mathbf{k}$=(0.5 0 0.25)), and a ferroelectric phase with incommensurate order where $\mathbf{P}$ is much reduced. Induced order of the R moments has been observed by scattering techniques in all phases, followed below $\sim$ 10K by a proper ordering that has however limited impact on the electrical properties.

In the present article, we report on the detailed magnetic structure of $GdMn_2O_5$ and on the use of x-ray magnetic scattering to fully determine the magnetic structure in this compound. This work is also motivated by the absence of spin-orbit coupling at first-order for Gd^{3+} ($4f^7$ electronic configuration) and its larger ionic radius with respect to known members (Tb, Ho, Er, Dy, Tm) and by the reported observation of ferroelectric behaviour together with a large spontaneous polarization (P_b >3600 $\mu C/m^2$)[8, 9] at low temperature. By combining x-ray magnetic scattering (XMS) data, collected off-resonance and at the Gd L_{III}-edge, we derive the complete magnetic phase diagram. We show that an incommensurate phase with wave-vector $\mathbf{k}$=(0.486,0,0.18) below T_{N1} ~40K uniquely collapses to a commensurate phase $\mathbf{k}$=($\frac{1}{2}$,0,0) below T_{N2} ~33K, without any further change of wave-vector below T_{N2}. Unlike any compounds of the same series, magnetic ordering of the Gd site follows a proper order parameter below T_{N2} suggesting that this ion is actively involved in the radical change of $\mathbf{k}$. From the quantitative analysis of the azimuthal dependence of the XMS signal, collected off-resonance and at the Gd L_{III}-edge, we propose a model of the Gd and Mn ordering in the commensurate phase and are able to explain why this specific magnetic configuration will induce a much larger electric polarization than for any related compounds. In addition we report the investigation of the magnetic domain topology of $GdMn_2O_5$ as function of electric field. We show maps of the crystal's surface in the ac plane on electric field switching along the b crystallographic direction. These maps evidence the evolution of the magneto-electric domains as function of the applied electric field.

2. Experimental

A high quality single crystal (1.2x0.9x0.5mm^3) of $GdMn_2O_5$ grown using floating zone method[10] was used for the x-ray magnetic scattering experiment, performed at the I16 beamline (Diamond Light Source Ltd.)[11, 12]. The sample was mounted on the beamline Kappa-diffractometer in a closed circle refrigerator with base temperature of $\sim$ 5K. The crystal was oriented with the diffraction face perpendicular to the (110) direction (Fig. 1). The azimuthal angle reference (ψ=0) is defined when the (100) reflection is in the scattering plane and its projection onto the incoming beam parallel to it. The diffractometer was operated in the vertical-plane scattering mode with an azimuth setup to allow for a 360° sample rotation about the scattering vectors. In this geometry the natural polarization of the incident beam is perpendicular to the scattering plane (σ). The integrated intensity of the reflections was measured using either a Dectris Pilatus 100K area detector and/or an avalanche photodiode point detector. In the latter case, the polarization of the reflected beam was linearly analyzed by rotating the scattering plane of a mosaic crystal chosen to have a Bragg angle close to 90° at the energy of interest (Al (220) at 6.4 keV, and Au (222) at 7.2455 keV). The measured Bragg intensities have been corrected for self-absorption and Lorentz factors, multiplying them by the factor $A(\mathbf{Q},\psi)\cdot L(\mathbf{Q}) = \left[1+\frac{\sin\alpha(\mathbf{Q},\psi)}{\sin\beta(\mathbf{Q},\psi)}\right]\cdot\sin 2\theta(\mathbf{Q})$ where $\alpha(\mathbf{Q},\psi)$ and $\beta(\mathbf{Q},\psi)$ are the incident and exit angles with respect to the crystal surface. To separate the contribution of the Gd from the Mn one, the photon energy was tuned close to the E_{res}(Gd L_{III}-edge)=7.2455 keV. In this condition

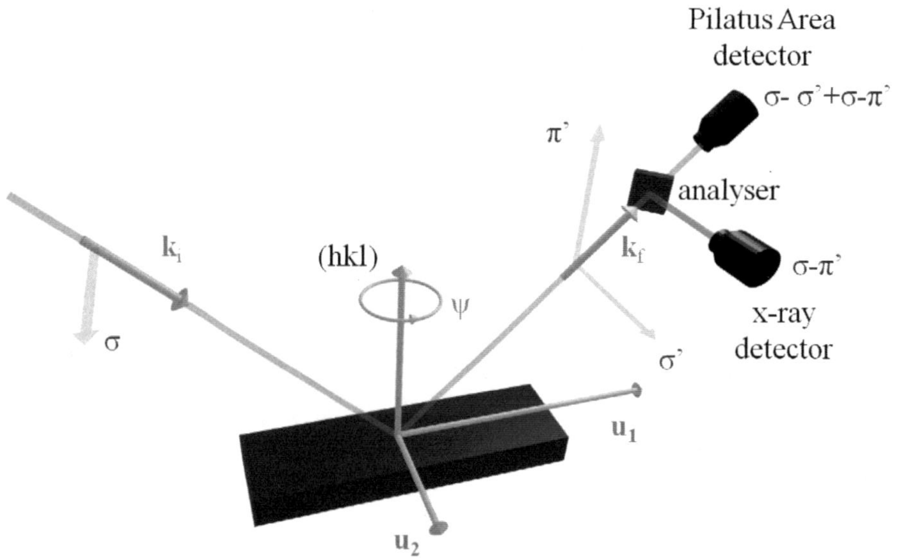

Figure 1. (Color online) A schematic diagram of the experimental layout. Sample mounted with the (110) surface cut in diffraction conditions.

a very large (factor ~40) enhancement (Fig. 2) of the magnetic scattering was observed which allowed to ascribe the observed signal entirely to the Gd magnetic moments. A different experimental configuration has been used to collect magnetic domain maps and apply an electric field along the *b* direction. The sample was rotated and mounted on a suitable sample holder with the *ac* plane in scattering conditions. Diamond phase plates have been used to select circular right and circular left light polarization of the incident x-ray beam and the beam size of 195x32 μm^2 has been reduced, using slits, to a ~50x50 μm^2 beam footprint on the sample surface. A Cu220 crystal has been used to analyse the scattered light in off-resonant conditions at 6.4 keV.

3. Results and Discussion

$GdMn_2O_5$ crystallizes in the centrosymmetric *Pbam* space group. Manganese ions are found in two oxidation states, Mn^{4+} (Wyckoff position 4f, 0,0.5,0.2551) and Mn^{3+} (Wyckoff position 4h,0.4118,0.3518,0.5), respectively in octahedral and square-pyramidal environments. The octahedra share edges to form chains along the *c*-axis. In the *ab*-plane, Mn polyhedra share edges and corners to form layers that can be viewed as an alternate stacking along *b* of ziz-zag ...Mn^{4+}-Mn^{3+}-Mn^{3+}-Mn^{4+}... chains running along the *a*-axis. The Gd^{3+} cations form GdO_8 polyhedra which are located on position 4g (0.1395,0.1716,0) and separate the adjacent "Mn" *ab*-layers.

The sequence of magnetic phase transitions of $GdMn_2O_5$ as shown in Fig. 2, has been obtained by sampling a large area of reciprocal space using an area detector. Additional Bragg reflections appear below 40K (red triangles in fig. 2), and can all be indexed by the incommensurate propagation vector $\mathbf{k}_{ICM}$ ~(0.49 0 0.18). It is important to note that the observation of two peaks at (k_x,0,k_z) and (1-k_x,0,k_z) confirms that the structure is also incommensurate along x, albeit the deviation from 0.5 is small. Below T_{N2} ~33K, k locks at

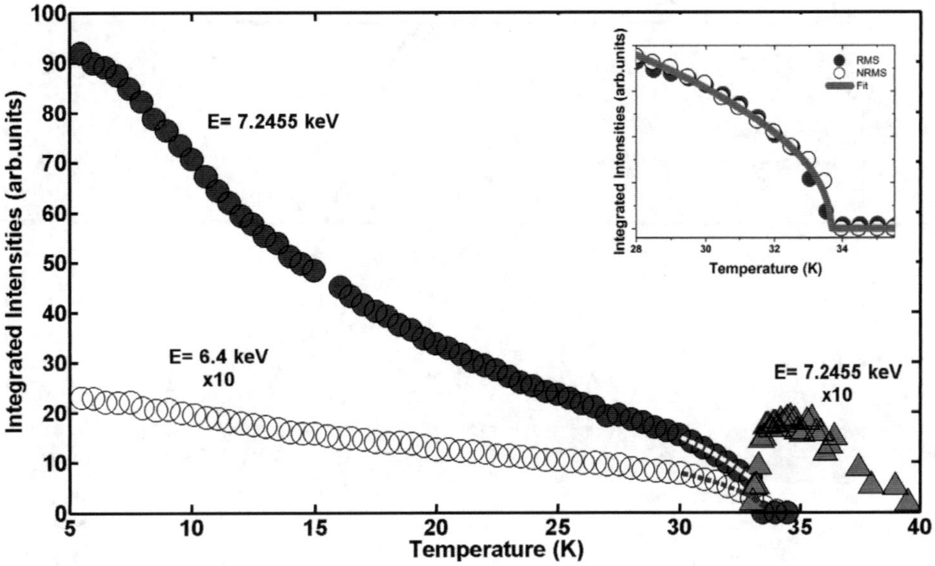

Figure 2. (Color online) Temperature dependence of the (2.5 3 0) reflection in resonant (full circles) and in off resonant (open circles) multiplied by a factor 10 conditions, collected at ψ=50 and ψ=0 respectively. Triangles represent the temperature dependence of the (4.49 4 0.18) reflection (x10), in resonant conditions. The dashed lines are fits to the critical exponents as described in the text. Error bars are contained in the spot size. Inset: normalised RMS and NRMS data of the (2.5 3 0) reflection and fit with a power law in the 28-35K region across the ICM-CM phase transition.

the commensurate value $\mathbf{k}_{CM}$=(1/2,0,0). In the commensurate phase in the vicinity of T_{N2}, the temperature dependence of the magnetic peaks intensities, shown in Fig.2, are adequately fitted with a power law (critical exponent 2β since the intensity is proportional to the square of the magnetization). The critical exponents measured in non resonant and resonant conditions are identical within the experimental error, respectively β=0.26±0.02 and β =0.29±0.03 [13]. This critical behavior indicates a unique order parameter with contribution from both the Gd and Mn magnetizations. This is in contrast to the observed induced magnetic ordering (secondary coupled order parameter) of Ho, Tb and Er observed in other compounds of the same family[14, 15].

The azimuthal dependence of five magnetic reflections measured in resonant conditions at T=5K (fig. 3(a)) has been used to derive a model for the magnetic structure of the Gd sublattice. In the dipolar approximation of the resonant case, the scattering is observed only in the rotated channel ($\sigma\pi'$) and the scattering amplitude f for a magnetic Bragg peak at momentum transfer $\mathbf{Q}$ and azimuth ψ is given by[16]:

$$f_{res}^{\sigma\pi'}(\mathbf{Q},\psi) \propto \mathbf{k'} \cdot \mathbf{S}_{Gd}(\mathbf{Q},\psi), \tag{1}$$

considering only the lowest order term, where $\mathbf{k'}$ is the scattered wave-vector and $\mathbf{S}_{Gd}$ the Gd magnetic structure factor. The technique probes indirectly the magnetic order of the f-electrons through the polarized Gd d density of states, as the d-f hybridization is allowed by the absence

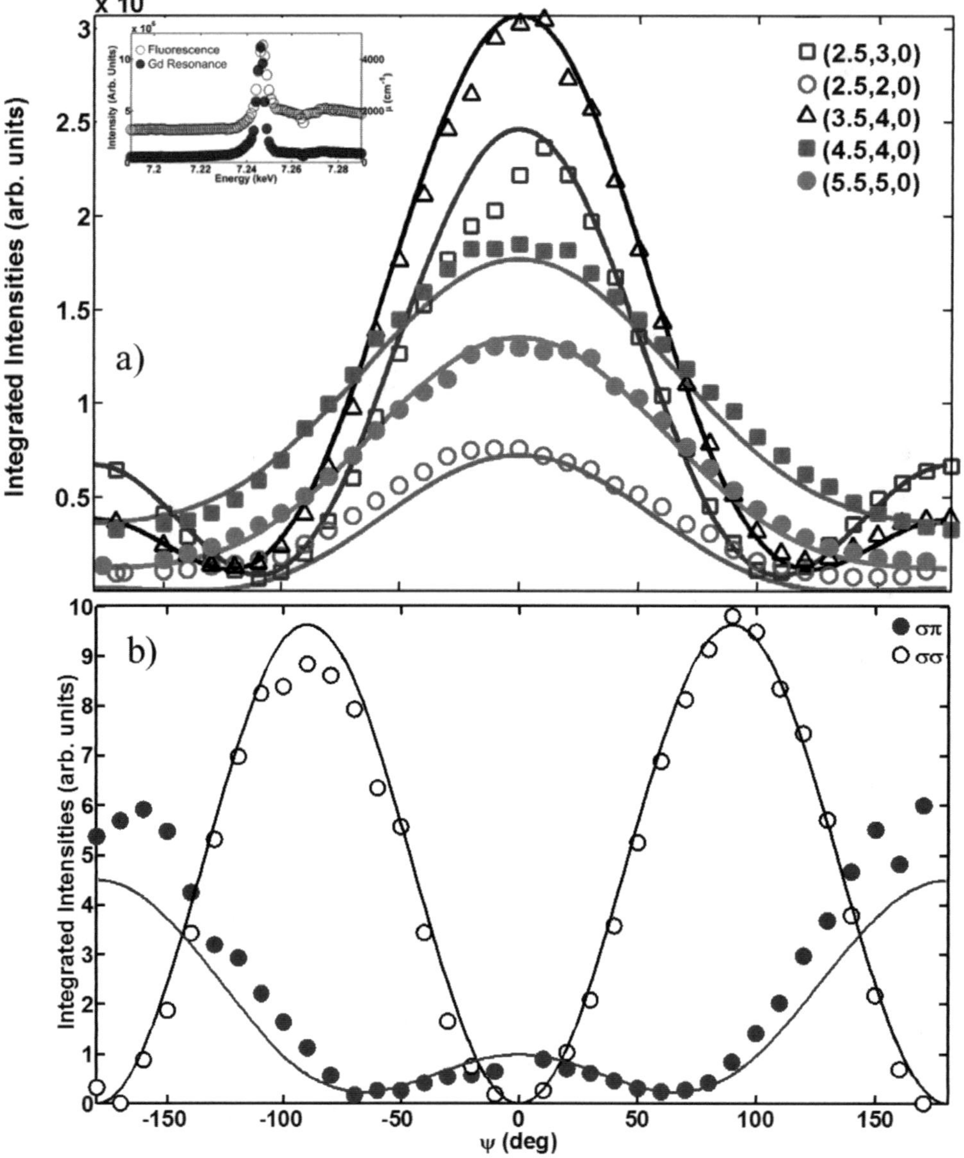

Figure 3. (Color online) Azimuthal dependence of the magnetic Bragg peak intensities at 5 K. The azimuth value is given with respect to a reference in the (1 0 0) direction. Upper panel: magnetic Bragg reflections at resonance (Gd L_{III} edge). The symbols containing the error bars show the experimental data points. The straight lines are fits to the data. Inset: Energy dependence of the magnetic reflection (2.5 3 0) (full circles) and of the fluorescence background (open circles) around the Gd L_{III} resonance energy. Lower panel: (2.5 3 0) reflection in non resonant condition at 6.4keV, in the $\sigma\sigma'$ (open circles) and $\sigma\pi'$ (full circles) channels. Error bars are contained in the spot size.

of inversion symmetry at the Gd site. The azimuthal scans present a two fold periodicity with maxima at positions close to ψ=0 and 180°, which given the azimuthal reference in the (100) direction, indicates that the Gd moments are approximately aligned along the crystallographic *a*-axis. The Gd sublattice magnetic configuration and its magnetic symmetry were determined empirically by least-square refinements of all azimuthal scans considered simultaneously. There is a unique solution for which the Gd moments (labelled 1-4 in Fig. 5) are oriented in the *ab*-plane and related by the time-reversed two-fold rotation axis along *b*.[1] A further refinement imposing this symmetry restriction lead to excellent results as shown in Fig. 3(a). For the magnetic structure stabilization, given the order parameter in the special direction (a,0), two irreducible representations (X_1,X_2) are allowed. The full magnetic space group found, which corresponds to X_2, is P_ab2_1a (P_aca2_1 in conventional International Tables for Crystallography settings). The proposed symmetry allowed magnetic mode (out of the six allowed ones spanning X_1 and X_2), is uniquely consistent with Gd moments in the *ab*-plane and a ferroelectric axis along *b*, as observed experimentally. With this symmetry, only the moments on site Gd1 and Gd2, on one hand, and Gd3 and Gd4, on the other, are related by the two-fold rotation, while the two sets are unrelated due to the loss of inversion symmetry. Finally, temperature dependent (T=5, 15, 25, 31K) energy scans of the resonant signal and of the azimuthal scans (not shown) suggested an unchanged Gd magnetic configuration.

The complete magnetic structure, Mn and Gd magnetic ordering and the relative phase between Gd and Mn modulations, was probed using non-resonant magnetic scattering (NRMS). In this case, the azimuthal dependence of the scattering amplitudes[16, 17, 18] in the $\sigma\sigma'$ and in the $\sigma\pi'$ channels used to fit the data are:

$$f^{\sigma\sigma'}_{nres}(\mathbf{Q},\psi) \propto (\mathbf{k}\times\mathbf{k'})\cdot\mathbf{S}(\mathbf{Q},\psi), \tag{2}$$

$$f^{\sigma\pi'}_{nres}(\mathbf{Q},\psi) \propto (1-\mathbf{k}\cdot\mathbf{k'})\mathbf{k}\cdot\mathbf{S}(\mathbf{Q},\psi) \tag{3}$$

where **k** is the incident wave-vector and **S** is the magnetic structure factor including Gd and Mn contributions. Due to the long collection time, only a single off-resonant azimuthal scan could be collected in the full azimuthal range, preventing to perform an unconstrained refinement including all Mn moments. Instead, the magnetic configuration of the Mn in the *ab* plane was fixed to that found for other commensurate structures of the RMn_2O_5 compounds, the main difference being caused by the different magnetic periodicity along *c*. As shown in Fig. 5, Mn^{4+}-Mn^{3+}-Mn^{3+}-Mn^{4+} form zig-zag AFM chains along the *a*-axis, and the moments are tilted by ~20° with respect to the crystallographic *a*-direction, pointing along the axis of the pyramidal Mn^{3+} site. Such configuration on its own does not account well for the observed off-resonance signal, in particular under-calculating the intensities in the $\sigma\pi'$ channel. However, by including the Gd contribution derived from the RMS work, the off-resonance scan can be adequately fitted, as shown in Fig. 3. Despite the fact that the relative orientation of the Mn moments with respect to the Gd moments could be inferred from the NRMS data since the total magnetic structure factor is sensitive to the relative phase, the lack of available data (only one scanned reflection) does not allow to uniquely discern between a ferromagnetic or antiferromagnetic Gd-Mn alignment. Preliminary neutron scattering performed on a recently isotropically substituted $GdMn_2O_5$ crystal confirms, as reported in Fig. 5, the Gd moments are arranged almost antiferromagnetically relative to the neighboring Mn^{3+} moments (pyramidal sites). By supposing that the Mn ordered moments are saturated at the spin-expected value of $3\mu_B$ (octahedral site, S=3/2) and $4\mu_B$ (pyramidal site, S=2), one obtains, from the NRMS data, an ordered moment for Gd^{3+} at 5K of $\sim 5.14(4)\mu_B$ and $\sim 4.75(4)\mu_B$ for Gd1/Gd2 and Gd3/Gd4 respectively. The Gd ordered moment is much larger than that found for R=Tb or

[1] Note that since the wave-vector doubles the magnetic unit-cell along *b*, an equivalent solution relates all these moments by a two-fold axis (without time reversal).

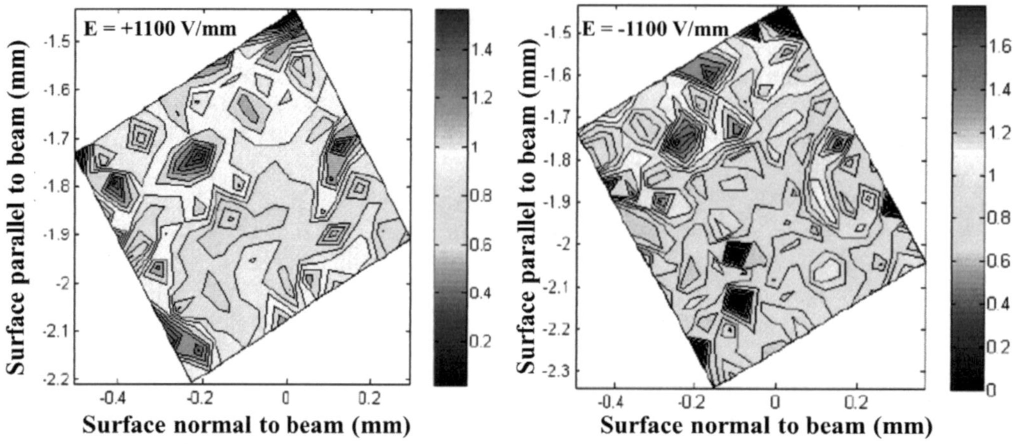

Figure 4. (Color online) Ratio between circular right and circular left light of the (0.5, 0, 2) magnetic reflection in different constant electric field (cooling) conditions: a) E = 1100 V/mm and b) E = -1100 V/mm. Colour maps representing inversion symmetry related magnetic domains (respectively ratio > 1 red, ratio < 1 blue). Pixels size is 50x50 μm^2 and the temperature is 5 K. Colours show the ratio between right and left circularly polarised light, corresponding to two different magnetic domain populations[19].

Ho.

In order to observe the magnetic domains and their evolution upon application of an external electric field, the sample was rotated and electrodes were applied close to the crystallographic *b* direction. The sample has been aligned with the [0.5, 0, 2] reflection in scattering conditions and it has been cooled to 5 K, with electric fields of + and - 1100 V/mm. The *ac* surface have been mapped by using incident circularly polarised light and by analysing the polarization of the scattered signal[19]. The ratio between circular right and circular left light has been measured in non-resonant conditions and plotted as function of the position on the sample surface. The resulting magnetic domain maps are shown in figure 4(a) and 4(b)). The data show that the magnetic domain populations was partially switched with the reversal of electric field but not fully reversed. We believe the combination of $\sim 25°$ tilt of the sample *b* axis with respect to the electric field direction and the maximum electric field generated by the Keithley electrometer used during this experiment wasn't enough to fully control the magnetic domain population in this compound. The domains size appear to be of the order of $\sim$ 100 microns and the larger ones (top left and centre right of fig 4, appear to act as pinning domains, blocking the domain population changes in the nearby regions.

4. Conclusions

The magnetic structure stable below T_{N2} of $GdMn_2O_5$ is the simplest of the series, and yet supports ferroelectricity along the *b* crystallographic axis. Although $GdMn_2O_5$ shows similarities with respect to other RMn_2O_5 compounds in the most polar phase, such as a locking of the propagation vector at half along the x direction, with the formation of AFM chains along the *a*-axis with constant moment amplitude, this is the only system for which k_z=0. The direct consequence of this periodicity is the production of a FM stacking along *c* of the adjacent AFM planes. This effect can be explained by two phenomena: firstly, eight-coordinated Gd^{3+} has a slightly larger ionic size (1.193 Å) than other rare-earth (Tb-1.18Å, Dy=1.16 Å, Ho=1.15 Å, Er=1.14 Å), which affects the Mn^{4+}-Mn^{4+} direct exchange interaction through the Gd layer. Since longer Mn-Mn inter-atomic distances promotes direct exchange (in manganese metals the crossover between AFM and FM is ~2.83Å), the argument seems to hold for $GdMn_2O_5$, where the Mn-Mn distance is 2.89Å. There is however a notable exception for $BiMn_2O_5$ which, despite an even larger ionic radius on the R site, displays antiferromagnetic stacking ($k_z = 1/2$). Secondly, based on the absence of first-order orbital momentum on the Gd site ($4f^7$ electronic configuration), it can be assumed that magnetism on the Gd site is very isotropic, and that the Gd moment direction will simply align along the spin of its strongest interacting neighbor. Since the Gd moments are nearly collinear with the first neighbor Mn^{3+} spins (Gd-Mn distance ~3.303Å), the latter explanation is in excellent agreement with experimental observations. The situation is very different in analogues with non-quenched orbital momentum displaying non-collinear arrangements of the R moments[10]. Due to this simple commensurate structure, unlike in any other ferroelectric RMn_2O_5, in $GdMn_2O_5$ the rare earth exhibit a large ordered moment in every layer. This unique magnetic configuration has important consequences for the ferroelectric behavior: since the Gd spin configuration on its own breaks inversion symmetry, one expect a finite contribution to the polarization along b from the coupled polar ionic displacements allowed by symmetry on the Gd site and coordinated oxygens. In the presence of a large ordered moment on Gd, one can anticipate a large contribution to the total polarization, provided that all polar displacements (Mn,O,Gd) are in phase. In this scenario, the Gd-Mn symmetric exchange striction, together with the Mn-Mn exchange striction mechanism, could lead to very large ferroelectric polarizations, as experimentally observed [9]. Ab-initio calculations, that will be facilitated by the small magnetic unit-cell, should shed light on the respective contributions of different ions to the total electric polarization, and separate the part imputable to the Gd magnetic ordering.

Acknowledgments

The work at Rutgers was supported by DOE grant DE-FG02-07ER46382.

References

[1] Kimura T, Goto T, Shintani H, Ishizaka K, Arima T and Tokura Y 2003 *Nature* **426** 55
[2] Radaelli P G, Chapon L C, Daoud-Aladine A, Vecchini C, Brown P J, Chatterji T, Park S and Cheong S W 2008 *Phys. Rev. Lett.* **101** 067205
[3] Chapon L C, Radaelli P G, Blake G R, Park S and Cheong S W 2006 *Phys. Rev. Lett.* **96** 097601
[4] Choi Y J, Yi H T, Lee S, Huang Q, Kiryukhin V and Cheong S W 2008 *Phys. Rev. Lett.* **100** 047601
[5] Sergienko I A, Şen C and Dagotto E 2006 *Phys. Rev. Lett.* **97** 227204
[6] Hur N, Park S, Sharma P A, Ahn J S, Guha S and Cheong S W 2004 *Nature* **429** 392
[7] Blake G R, Chapon L C, Radaelli P G, Park S, Hur N, Cheong S W and Rodríguez-Carvajal J 2005 *Phys. Rev. B* **71** 214402
[8] Inomata A and Kohn K 1996 *Journal of Physics: Condensed Matter* **8** 2673 URL `http://stacks.iop.org/0953-8984/8/i=15/a=016`
[9] Lee N, Vecchini C, Choi Y J, Chapon L C, Bombardi A, Radaelli P G and Cheong S W 2013 *Phys. Rev. Lett.* **110**(13) 137203 URL `http://link.aps.org/doi/10.1103/PhysRevLett.110.137203`
[10] Vecchini C, Chapon L C, Brown P J, Chatterji T, Park S, Cheong S W and Radaelli P G 2008 *Phys. Rev. B* **77** 134434
[11] *http://www.diamond.ac.uk/Home/Beamlines/I16.html* URL `http://www.diamond.ac.uk/Home/Beamlines/I16.html`
[12] Collins S P, Laundy D and Stunault A 2001 *Journal of Physics: Condensed Matter* **13** 1891 URL `http://stacks.iop.org/0953-8984/13/i=9/a=312`
[13] Toledano J and Toledano P 1987 *The Landau Theory of Phase Transitions* (World Scientific, Singapore)
[14] Beutier G, Bombardi A, Vecchini C, Radaelli P G, Park S, Cheong S W and Chapon L C 2008 *Phys. Rev. B* **77** 172408

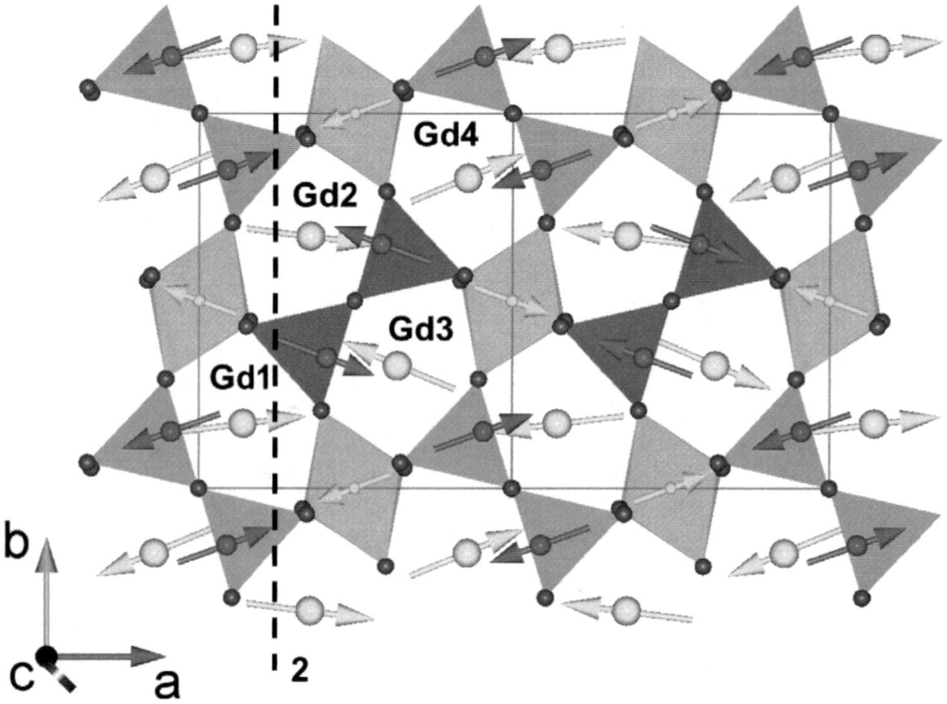

Figure 5. (Color online) Magnetic structure of $GdMn_2O_5$ projected in the *ab* plane. The structure is shown in two unit-cells, marked by thin black lines, along the *a*- and *b*-axis. The green (small and light grey), red (grey) and blue (black) arrows represent magnetic moments on Mn^{4+}, Mn^{3+} and Gd^{3+} sites respectively. Corresponding Mn-O polyhedra are shown with the same colors.

[15] Kimura H, Kobayashi S, Fukuda Y, Osawa T, Kamada Y, Noda Y, Kagomiya I and Kohn K 2007 *Journal of the Physical Society of Japan* **76** 074706 URL http://jpsj.ipap.jp/link?JPSJ/76/074706/
[16] Blume M and Gibbs D 1988 *Phys. Rev. B* **37** 1779–1789
[17] Hill J P and McMorrow D F 1996 *Acta Crystallographica Section A* **52** 236–244 URL http://dx.doi.org/10.1107/S0108767395012670
[18] De Bergevin F and Brunel M 1981 *Acta Cryst.* **A37** 314
[19] Fabrizi F, Walker H C, Paolasini L, de Bergevin F, Boothroyd A T, Prabhakaran D and McMorrow D F 2009 *Phys. Rev. Lett.* **102**(23) 237205 URL http://link.aps.org/doi/10.1103/PhysRevLett.102.237205

Existence of covalent-like bonding in $Pd_{40}Cu_{20}Ge_{40}$ metallic glass observed by anomalous x-ray scattering

T Yamamoto[1], S Hosokawa[2,3], J-F Bérar[4], N Boudet[4], W-C Pilgrim[3], K Hayashi[5], H Kato[5]

[1]Department of Mechanical and Intelligent Engineering, Utsunomiya University, Utsunomiya 321-8585, Japan
[2]Department of Physics, Graduate School of Science and Technology, Kumamoto University, Kumamoto 860-8555, Japan
[3]Department of Chemistry, Physical Chemistry, Philipps University of Marburg, D-35032 Marburg, Germany
[4]Institut Néel, Centre National de la Recherche Scientifique/Université Joseph Fourier (CNRS/UJF), 38042 Grenoble Cedex 9, France
[5]Institute of Materials Research, Tohoku University, Sendai 980-8577, Japan

E-mail: hosokawa@sci.kumamoto-u.ac.jp

Abstract. To elucidate the local structure of $Pd_{40}Cu_{20}Ge_{40}$ metallic glass having curious thermodynamic properties of a low melting point and an extremely small ratio of glass transition- and melting temperatures, anomalous x-ray scattering experiments were performed close to the Pd, Cu, and Ge K edges. The results were analyzed using reverse Monte Carlo modeling to obtain partial structure factors and pair correlation functions. Covalent-like bondings between the Pd and Ge atoms are realized in this metallic glass, which may be related to the thermodynamic anomalies.

1. Introduction

The ratio of glass transition- and melting temperatures, T_g/T_m, is widely employed as a simple index to characterize glass-forming abilities of glasses [1]. Generally, values of 3/2 or at smallest 1/2 were observed in various types of glasses, such as oxide, organic, and metallic glasses. Exceptions are only observed in simple polymer glasses, such as polyethylene (0.35) and polytetrafluoroethylene (0.26) [2, 3]. Recently, Yamamoto *et al.* [4] have found a curious metallic glass, $Pd_{40}Cu_{20}Ge_{40}$, having a very low T_g and an extremely small T_g/T_m of 0.36, where $T_g = 52°C$ and $T_m = 625°C$. Surprisingly, this metallic glass is formable even in hot water.

To investigate an origin of such a small T_g/T_m value in $Pd_{40}Cu_{20}Ge_{40}$ metallic glass from a partial structural point of view, we have measured anomalous x-ray scattering (AXS) close to the Pd, Cu, and Ge K x-ray absorption edges. The obtained differential structure factors, $\Delta_k S(Q)$, were analyzed using reverse Monte Carlo (RMC) modeling to clarify local- and intermediate-range partial correlations between the constituent elements. In this paper, we report our finding of clusters of Pd and Ge atoms with a highly covalent-like nature.

Journal of Physics: Conference Series **519** (2014) 012005 doi:10.1088/1742-6596/519/1/012005

2. Experimental procedure and data analysis

A $Pd_{40}Cu_{20}Ge_{40}$ metallic glass sample was manufactured by a normal melt-spinning technique with a tapered silica nozzle. The glass phase of the obtained ribbons was examined by x-ray diffraction and transmission electron microscopy. The composition of the ribbons was confirmed to be nominal value by an electron-probe micro-analysis equipment.

AXS experiments were carried out at room temperature using a standard $\omega-2\theta$ diffractometer installed at the beamline BM02 of the European Synchrotron Radiation Facility (ESRF) in Grenoble, France. Scattering measurements were performed at two incident x-ray energies below each K edge, i.e., –30 eV for Pd, –20eV for Cu and Ge, and –200 eV for all elements. For discriminating elastic signals from $K\beta$ fluorescence- and Compton scattering contributions, a bent graphite crystal analyzer was mounted at the middle of a 1-m-long detector arm. An energy resolution was about 120 eV near the Pd K edge and about 50 eV near the Cu and Ge K edges, with which the $K\beta$ contribution is estimated to be less than 0.1%. More than three or five million counts were collected at the maximum of the structure factor, $S(Q)$, within a reasonable beamtime for this AXS experiment. The feasibility of this detecting system is given elsewhere [5]. For AXS data analysis, Sasaki's theoretical values [6] were used for the anomalous terms together with theoretical usual terms [7]. Following the procedure given in Ref. [8], $\Delta_k S(Q)$'s were calculated for each K edge.

RMC modeling [9] is a useful tool to construct three-dimensional atomic configurations of disordered materials using only experimental diffraction data. In the RMC calculations, atoms are moved from an initial configuration so as to minimize deviations from experimental data, e.g., in this study, three $\Delta_k S(Q)$'s and $S(Q)$ measured near the Pd K edge, using a standard Metropolis Monte Carlo algorithm [10].

A starting configuration of a system containing a total of 10,000 atoms with the measured number density of 69.86 nm^{-3} was generated using a hard-sphere Monte Carlo simulation (a random configuration except short bonds). To avoid unphysical atomic configurations, only a constraint of shortest atomic distances was applied to the RMC calculations, i.e., 0.19, 0.21, and 0.22 nm for Ge-Ge, (Pd,Cu)-Ge, and other inter-metallic (Pd,Cu) correlations, respectively. The RMC calculations were performed using the RMC++ program package coded by Gereben *et al.* [11].

3. Results and discussion

Circles in Fig. 1 show $\Delta_k S(Q)$'s close to the Pd, Cu, and Ge, together with $S(Q)$, obtained from the present AXS measurements. The limited Q ranges in the $\Delta_{Cu} S(Q)$ and $\Delta_{Ge} S(Q)$ are due to the low incident energies near the Cu and Ge K absorption edges. Of special interest is the spectral shape of $\Delta_{Ge} S(Q)$, i.e., there is a steep minimum at 19 nm^{-1}, and the first peak at about 28 nm^{-1} is not high. These features look far from those of usual metallic glasses, and are similar to those of covalent chalcogenide glasses [8].

The $\Delta_{Pd} S(Q)$ and $S(Q)$ functions have small shoulders at about 20 nm^{-1}. Since the shoulder in $\Delta_{Pd} S(Q)$ is more prominent than that in $S(Q)$, and no clear indications are seen in $\Delta_{Cu} S(Q)$ and $\Delta_{Ge} S(Q)$, the origin of the shoulder seems to be Pd-related partial correlations. The $\Delta_{Cu} S(Q)$ function shows an enhancement in the low Q range, and thus, a density or concentration fluctuation is expected around the Cu atoms.

Solid curves in Fig. 1 show the best fits of the RMC modeling. Although the $\Delta_k S(Q)$ functions have features very different from each other as mentioned above, they surprisingly coincide well with the corresponding experimental diffraction data.

Figure 2(a) shows partial structure factors, $S_{ij}(Q)$, obtained from the present RMC modeling. As seen in the figure, large prepeaks are realized at about 19 nm^{-1} in all of the homopolar correlations of $S_{PdPd}(Q)$, $S_{CuCu}(Q)$, and $S_{GeGe}(Q)$. It should be noted that only a small and broad shoulder was observed in Pd- [12] and Zr-based [13] metallic glasses so far, and this is the

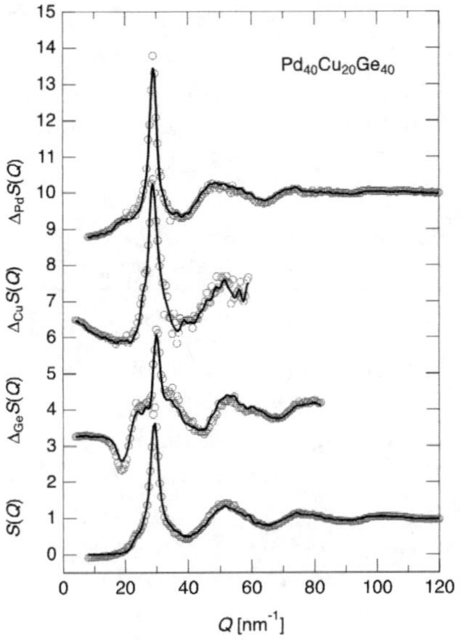

Figure 1. (Color online.) From top to bottom: $\Delta_k S(Q)$'s close to the Pd, Cu, and Ge K edges, and $S(Q)$. Circles indicate those obtained from the AXS experiments, and solid curves denote best fits of the RMC modeling. For clarity, the spectra are displaced upwards by three each.

first finding of such a large prepeak in metallic glasses.

On the contrary, the heteropolar $S_{\mathrm{PdGe}}(Q)$ function has a large and negative minimum at the Q position of the prepeaks, indicating very strong Pd-Ge heteropolar correlations analogous to chalcogenide glasses [8]. However, the remaining Cu-related heteropolar correlations, Pd-Cu and Cu-Ge, show features similar to usual metallic glasses indicating metallic bondings around the Cu atoms. Only the $S_{\mathrm{PdCu}}(Q)$ function indicates a weak concentration fluctuation tendency in the low Q region below 18 nm^{-1}.

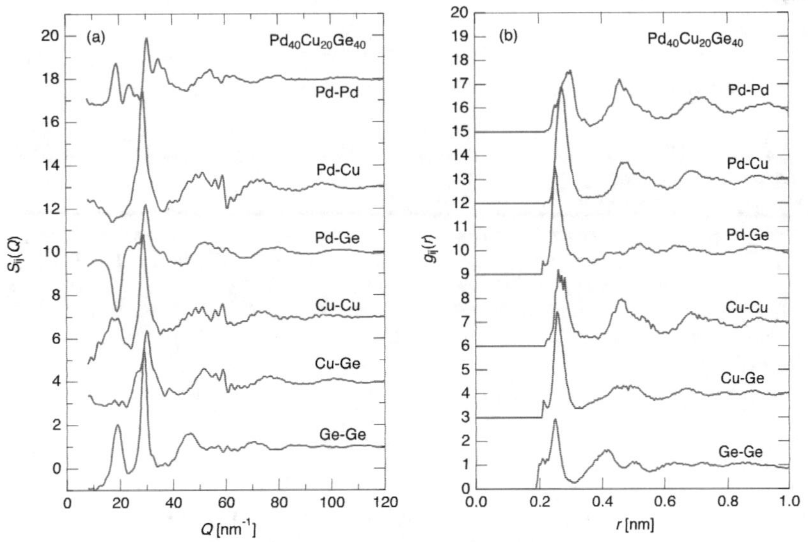

Figure 2. (Color online.) (a) $S_{ij}(Q)$ and (b) $g_{ij}(r)$ functions obtained from the RMC fits. For clarity, the spectra are displaced upwards.

Figure 2(b) shows partial pair distribution functions, $g_{ij}(r)$, obtained from the RMC modeling. At a glance, all of the $g_{ij}(r)$ functions seem to be usual as of metallic glasses, i.e., dense-packed features with a large first peak followed by damping oscillations. However, the first peaks in the homopolar $g_{ij}(r)$ functions are small in height compared with the corresponding second peaks, which is the reason of the existence of prepeaks in the corresponding $S_{ij}(Q)$ spectra.

The first coordination shells were defined up to mostly the first minima of the $g_{ij}(r)$ functions, i.e., 0.35 nm around the Pd atoms and 0.33 nm around the Cu and Ge atoms. The total coordination numbers around the Pd, Cu, and Ge atoms are about 13.0, 12.5, and 10.7, respectively, which seem to indicate the dense-packed features of 12. As expected from the features of the $g_{ij}(r)$ functions, however, the portions of homopolar partial coordinations of Pd-Pd, Cu-Cu, and Ge-Ge pairs are 35.8%, 16.4%, and 31.8% , respectively, which are much smaller than the nominal values of 40%, 20%, and 40%, in particular, Ge-Ge exceeding 20% of the discrepancy. Thus, the heteropolar correlations are much favorable in the first shells of this metallic glass.

From these analyses, we found a heterogeneous nature of the structure of $Pd_{40}Cu_{20}Ge_{40}$ metallic glass. The Pd and Ge atoms are strongly bound to form covalent-like clusters, while the Cu atoms seem to connect with the Pd and Ge atoms in a manner of weak metallic bonding. In relation to the small small $T_g/T_m = 0.36$ value in this metallic glass, these structural findings suggest the existence of covalent-like bondings between the Pd and Ge atoms.

4. Conclusion

AXS measurements were performed close to the Pd, Cu, and Ge K edges on $Pd_{40}Cu_{20}Ge_{40}$ metallic glass to search the local structural origin of curious thermodynamic properties of a low $T_g = 52°C$ and an extremely small $T_g/T_m = 0.36$. The results were analyzed using reverse Monte Carlo modeling to obtain $S_{ij}(Q)$'s and $g_{ij}(r)$'s. The existence of covalent semiconductor-like configurations between the Pd and Ge atoms may be related to the thermodynamic anomalies in the $Pd_{40}Cu_{20}Ge_{40}$ metallic glass.

Acknowledgments

We acknowledge Dr. Kohara of JASRI/SPring-8 for the support of RMC calculations. The AXS experiments were performed at BM02/ESRF (Proposal No. HD-426).

References

[1] Inoue A 2000 *Acta Mater.* **48** 279
[2] Andrews R J and Grulke E A 1999 *Glass Transition Temperatures of Polymers*, in Brandrup J, Immergut E H and Grulke E A (Eds.), *Polymer Handbook 4th Ed.* (John Wiley & Sons, Inc., New York) p. VI/193-277
[3] R. L. Miller 1999 *Crystallographic Data and Melting Points for Various Polymers*, in Brandrup J, Immergut E H and Grulke E A (Eds.), *Polymer Handbook 4th Ed.* (John Wiley & Sons, Inc., New York) p. VI/1-192
[4] Yamamoto T, Hayashi K, Seki I, Suzuki K and Ito M 2013 *J. Non-Cryst. Solids* **359** 46
[5] Hosokawa S and Bérar J-F 2007 *AIP Conf. Proc.* **879** 1743
[6] Sasaki S 1989 *KEK Report 1989* (Nat. Lab. High Energy Phys., Tsukuba) p. 1
[7] MacGillavry C H and Rieck G D (ed.) 1969 *International Table for X-ray Crystallographie, 2nd ed.* (Kynoch, Birmningham) Vol. III
[8] Hosokawa S, Oh I, Sakurai M, Pilgrim W-C, Boudet N, Bérar J-F and Kohara S 2011 *Phys. Rev. B* **84** 014201, and references therein
[9] McGreevy R L and Pusztai L 1988 *Mol. Simul.* **1** 359
[10] Metropolis N, Rosenbluth A W, Rosenbluth M N, Teller A. H and Teller E 1953 *J. Phys. Chem.* **21** 1087
[11] Gereben O, Jóvári P, Temleitner L and Pusztai L 2007 *J. Optelectr. Adv. Mater.* **9** 3021
[12] Hosokawa S, Sato H, Ichitsubo T, Nakatake M, Happo N, Bérar J-F, Boudet N, Usuki T, Pilgrim W-C, Matsubara E and Nishiyama N 2009 *Phys. Rev. B* **80** 174204
[13] Hosokawa S, Bérar J-F, Boudet N, Kohara S, Pilgrim W-C, Joester M, Stellhorn J, Zeidler A, Maruyama K, Nasu T, Zhang W and Kato H 2012 *Int. J. Mater. Res.* **103** 1108

Phase transition of KDP observed by Resonant X-ray Diffraction at forbidden reflections

G. Beutier[1,2], S. P. Collins[2], E. N. Ovchinnikova[3], G. Nisbet[2], V. E. Dmitrienko[4]

[1] SIMaP, Grenoble Institute of Technology - CNRS - Grenoble Alpes University, France
[2] Diamond Light Source, Harwell Science & Innovation Campus, OX11 0DE, United Kingdom.
[3] Department of Physics of Moscow State University, 119899 Moscow, Russia
[4] A.V. Shubnikov Institute of Crystallography, 119333, Moscow, Russia

E-mail: guillaume.beutier@simap.grenoble-inp.fr

Abstract. We report observations of space-group-forbidden Bragg reflections in Potassium Dihydrogen Phosphate (KH_2PO_4), also known as KDP, measured by resonant x-ray diffraction at the potassium K edge. We find clear evidence for a transition from one class of space-group-forbidden reflections, where scattering is ruled out by the electric dipole approximation, to a second class, in the ferroelectric phase, where scattering can proceed due to resonant anisotropy within the dipole approximation. The change of symmetry is clearly evidenced by the sudden change of intensity and energy spectrum of the forbidden reflections.

1. Introduction

Diffraction of x-rays, neutrons and electrons is a very sensitive technique for investigations of structural phase transitions because the reduction in symmetry that drives the transition is often heralded by the arrival of weak diffraction peaks that become allowed due to the change in the space group extinction rules. Such a transition from forbidden to allowed scattering can be considered as a special case of a more general hierarchy of scattering processes that involve various ranks of the scattering tensor - in this case, scalar (rank-zero) scattering. When scalar scattering (including Thomson scattering) is forbidden, the crystal symmetry may allow resonant scattering to proceed via anisotropy in the resonant scattering within the electric dipole approximation, described by a second-rank tensor. As symmetry increases, even this process may be annihilated, while allowing higher-rank processes to occur due to some exotic scattering mechanism, described by tensors of rank three or four.

Potassium Dihydrogen Phosphate (KH_2PO_4), also known as KDP is a fascinating test-bed for these processes: the ferroelectric phase transition provides an example of a transition from forbidden- to super-forbidden scattering. Our observations of resonant scattering from KDP, at the potassium K-edge, provide a clear demonstration of this phenomenon.

KDP, and many of its isomorphs, are important hydrogen-bonded ferroelectrics whose properties have long been a source of interest. It spans a wide range of room temperature technological applications, from food additives [1] to optical modulators in non-linear optics [2]. From the physical point of view, it has been mostly studied for its ferroelectricity, being one of the first discovered ferroelectric materials [3]. Indeed, ferroelectricity in KDP arises below

Journal of Physics: Conference Series **519** (2014) 012006 doi:10.1088/1742-6596/519/1/012006

122 K in a complex mechanism that has long been debated. This ferroelectric phase transition has triggered a lot of theoretical advances, based on the pseudospin model initially proposed by Slater [4] and later developed by many authors [5]. It is also an example of soft phonon mode driven phase transition [6]. Experimentally, KDP has been extensively studied against temperature [7], pressure [8], deuteration [9], electric field [10] and mechanical stress [11]. The mechanism is the following: The crystal structure is made of PO_4 tetrahedra linked to each other by hydrogen bonds. The latter consists of individual protons tunnelling between two equivalent quantum wells either side of the mid-point between the tetrahedra. In the paraelectric phase, they tunnel randomly between both sites, whereas in the ferroelectric phase they freeze in a long range ordered pattern. This pattern distorts the structure such that the ionic displacements of the Potassium and Phosphorus ions induces an electric polarisation [12]. The transition is hence at the same time an order-disorder transition for the protons and a displacive transition for the other ions. In the following, we will refer to the ferroelectric phase as the ordered phase and to the paraelectric phase as the disordered phase, since this is the structural detail that matters for the forbidden reflections. After a long controversy, it has been established that the phase transition is of the first order but nearly second order [13].

In this paper we use this now rather well understood phase transition to generalize the concept of forbidden reflections. We will show that the subtle change of symmetry occurring at the phase transition can enable or disable some of the resonant terms involved in forbidden reflections.

2. Theoretical

2.1. Crystal symmetries and forbidden reflections

In the disordered phase, KDP has a tetragonal crystal structure with space group $I\bar{4}2d$, which becomes orthorhombic with space group $Fdd2$ in the ordered phase [14]. The electric polarization of the ordered phase appears along the unique axis of the tetragonal structure. In the following, the settings of the space group $Fdd2$ will be used to describe the crystal. The tetragonal structure will be treated as a special case, since $Fdd2$ is a subgroup of $F\bar{4}d2$, which is another representation of $I\bar{4}2d$. In fact, the description of the disordered phase in the $I\bar{4}2d$ space group corresponds to the average structure, in which the protons have equal probabilities to be in either minima of the double well, but the instantaneous unit cell symmetry does not have such high symmetry. The transformation from the $I\bar{4}2d$ structure to the $F\bar{4}d2$ structure corresponds to a rotation of 45° around the c-axis, which automatically induces a doubling of the unit cell.

In either crystallographic phase, KDP has two types of forbidden reflections: first, the reflections of the type hkl, with h, k, l not all of the same parity, are forbidden by the centring symmetry; secondly, the reflections of the type $h0l$ with $h+l \neq 4n$ and $0kl$ with $k+l \neq 4n$ are forbidden by glide-plane symmetries. The centring symmetry and the glide-plane symmetries survive the phase transition, such that reflections forbidden in one phase remain forbidden in the other phase, preventing an easy observation of the phase transition. This rule is true for scalar atomic scattering factors, but resonant x-rays enhance higher rank tensors which can have non-vanishing structure factors. The reflections forbidden by the centring symmetry are forbidden for all tensors of any rank: their structure factors vanish because of the pure translational symmetry. But in the case of glide-plane forbidden reflections, higher rank tensors can have non-vanishing structure factors. A detailed investigation of the tensor symmetries is necessary to determine which ones participate in the forbidden reflections. We will now focus on reflections that are forbidden by a glide-plane but not by the centring symmetry. They are of the type $h0l$ with h and l even and $h+l = 4n+2$, and $0kl$ with k and l even and $k+l = 4n+2$. The structure factor F for such reflections of the resonant atomic scattering tensor $\boldsymbol{f}$ is:

$$F(\boldsymbol{f}) = 4\left[\boldsymbol{f} - \hat{S}(\boldsymbol{f})\right] \quad (1)$$

where $\hat{S}$ is the symmetry operator associated with the glide-plane. Therefore, only those tensor components of $\boldsymbol{f}$ which are not invariant with respect to glide-plane symmetry contribute to the diffraction at these reflections. Since these reflections are forbidden, the scalar structure factor vanishes and only resonant atoms, which have enhanced higher rank tensors of scattering, give a significant contribution. In the present case, resonant atoms are the potassium atoms and occupy special positions $8a$ of space group $Fdd2$ (or equivalently at special position $4a$ of space group $I\bar{4}2d$). This crystallographic site has a two-fold axis along 001 (point symmetry ..2) in the orthorhombic phase, which becomes a four-fold rotoinversion axis (point symmetry $\bar{4}$..) in the tetragonal phase. We will see that the change of local symmetry from two-fold axis to four-fold rotoinversion axis is at the origin of drastic changes in the resonant reflections.

2.2. Resonant X-ray Diffraction

Resonant X-ray Diffraction (RXD) is the application to the diffraction geometry of a more general technique now referred as Resonant Elastic X-ray Scattering (REXS). By exciting an electronic transition of the sample, resonant x-rays acquire an enhanced sensitivity to the resonant species and its local environment. It is a powerful probe of electronic orders, including magnetism [15]. REXS is usually described in terms of electronic multipolar resonances. The pure dipole dipole (E1E1) contribution is normally the strongest, but it may cancel out in the structure factor of forbidden reflections, depending on the symmetry of the resonant site, the global symmetry of the crystal, and the considered reflection. Such case allows for easier observation of weaker resonant terms such as mixed dipole-quadrupole (E1E2) terms [16], pure quadrupole-quadrupole (E2E2) terms [17] and maybe higher order electric multipoles when these vanish too. It also allows for easier observation of weaker processes of E1E1 origin such as Thermal Motion Induced (TMI) scattering [18, 19] or Point-Defect Induced (PDI) scattering [18].

Such is the case for resonant forbidden reflections in the disordered phase of KDP at the potassium K edge: because of the $\bar{4}$ axis of the resonant site, the E1E1 tensor is uni-axial and along the 001 axis. It is therefore invariant with respect to the glide-planes of the space group, which all contain the 001 axis, and its structure factor cancels out at the corresponding forbidden reflections. The first non-vanishing tensors are thus third rank tensors. Conversely, the Potassium site in the ordered phase has only a two-fold axis, allowing for additional components of the E1E1 tensor which do not cancel out in the structure factor. The appearance and disappearance, across the phase transition, of the E1E1 term in the structure factor of glide-plane forbidden reflections, has a drastic impact on their intensity and spectrum.

Let us now detail the tensors contributing to the atomic scattering factors and their structure factor at the considered forbidden reflections. We shall use the Cartesian formalism, which is convenient with orthorhombic systems, with axes x, y and z respectively along the 100, 010 and 001 of the orthorhombic structure. Within this framework, the multipolar expansion of REXS up to the electric quadrupole term has the following form [20]:

$$\boldsymbol{f} = \epsilon'^{*}_{\alpha}\epsilon_{\beta}\left[D_{\alpha\beta} + \frac{i}{2}(k_{\gamma}I_{\alpha\beta\gamma} - k'_{\gamma}I^{*}_{\beta\alpha\gamma}) + \frac{1}{4}k'_{\gamma}k_{\delta}Q_{\alpha\gamma\beta\delta}\right] \quad (2)$$

with $\{\alpha, \beta, \gamma, \delta\} \in \{x, y, z\}^4$ and summation over repeated indices is implied. The vectors $\boldsymbol{k}$, $\boldsymbol{k}'$ and $\boldsymbol{\epsilon}$, $\boldsymbol{\epsilon}'$ are the propagation and polarization vectors, respectively, of the incident and scattered radiations. Below we shall also use conventional $\boldsymbol{\sigma}$ and $\boldsymbol{\pi}$ polarization vectors, normal and parallel, respectively, to the scattering plane. Magnetic multipolar and higher order electric multipolar contributions, which are expected extremely weak [21], were omitted. Thus, the resonant scattering tensor consists only of the second rank E1E1 tensor $\boldsymbol{D}$, the third rank E1E2 tensor $\boldsymbol{I}$, and the fourth rank E2E2 tensor $\boldsymbol{Q}$, all of which possessing tensor elements that are sensitive to the incident radiation energy with their own spectrum.

The E1E1 tensor $\boldsymbol{D}$ is of rank 2 and is symmetric in the case of a non-magnetic crystal. Because of the (at least) 2-fold axis at the potassium site, it has the following form [22]:

$$\boldsymbol{D} = \begin{pmatrix} D_{xx} & D_{xy} & 0 \\ D_{xy} & D_{yy} & 0 \\ 0 & 0 & D_{zz} \end{pmatrix} \tag{3}$$

Following equation 1, its structure factor at the glide-plane forbidden reflections is:

$$F(\boldsymbol{D}) = 4\left[\boldsymbol{D} - \hat{S}(\boldsymbol{D})\right] = 8 \begin{pmatrix} 0 & D_{xy} & 0 \\ D_{xy} & 0 & 0 \\ 0 & 0 & 0 \end{pmatrix} \tag{4}$$

If D_{xy} is not zero for any symmetry reason and if the E1E1 term is considerably stronger than any other term, as it is usually the case, the spectrum of the forbidden reflection is unique, independently of the chosen forbidden reflection, its azimuth, and the beam polarization (after correction for self-absorption effects, taking into account linear dichroism [23]). We will show that it is the case in the ordered phase. In the disordered phase, the 2-fold axis becomes a pseudo 4-fold axis, and the off-diagonal terms of symmetric rank 2 tensors vanish [22]: $D_{xy} = 0$, thus cancelling out the structure factor: above the transition the reflection is forbidden for scattering tensors up to the rank 2, whereas below the transition it is forbidden only for rank 0 (scalars). This can be seen as an extension of the concept of forbidden reflections, usually understood as for scalar scattering factors.

In absence of E1E1 scattering, it is necessary to extend the analysis to weaker scattering processes represented by higher order tensors. E1E2 (third-rank tensor) and E2E2 (fourth-rank tensor) resonances are usually the strongest ones, along with the TMI term (third-rank tensor). PDI scattering is also expected to have a sizeable effect in the paraelectric phase of KDP, because of the static disorder of the protons [24]. Here we detail the analysis for the E1E2 term only, only to show that it does not vanish. The symmetry properties of third rank tensors representing TMI and PDI scattering differ slightly from that of the E1E2 term [25].

Any third rank Cartesian tensor $\boldsymbol{T}$ with $\bar{4}$ symmetry has six independent components [22, 26]: $T_{xxz} = -T_{yyz}$, $T_{zxx} = -T_{zyy}$, $T_{xzx} = -T_{yzy}$, $T_{xyz} = T_{yxz}$, $T_{xzy} = T_{yzx}$ and $T_{zxy} = T_{zyx}$. In the notation scheme of [22]:

$$\boldsymbol{T} = \left(\begin{array}{ccc|ccc|ccc} 0 & 0 & 0 & T_{xyz} & T_{xzx} & 0 & T_{xzy} & T_{xxz} & 0 \\ 0 & 0 & 0 & -T_{xxz} & T_{xzy} & 0 & -T_{xzx} & T_{xyz} & 0 \\ T_{zxx} & -T_{zxx} & 0 & 0 & 0 & T_{zxy} & 0 & 0 & T_{zxy} \end{array}\right) \tag{5}$$

By applying the glide-plane symmetry, the general structure factor for a third rank tensor is left with 3 independent components [27]:

$$F(\boldsymbol{T}) = 4\left[\boldsymbol{T} - \hat{S}(\boldsymbol{T})\right] = 8 \left(\begin{array}{ccc|ccc|ccc} 0 & 0 & 0 & T_{xyz} & 0 & 0 & T_{xzy} & 0 & 0 \\ 0 & 0 & 0 & 0 & T_{xzy} & 0 & 0 & T_{xyz} & 0 \\ 0 & 0 & 0 & 0 & 0 & T_{zxy} & 0 & 0 & T_{zxy} \end{array}\right) \tag{6}$$

Additionally, the E1E2 tensor $\boldsymbol{I}$ has the following property by construction [20]: $I_{\alpha\beta\gamma} = I_{\alpha\gamma\beta}$ for all $\{\alpha, \beta, \gamma\} \in \{x, y, z\}^3$. Its structure factor at the glide-plane forbidden reflections is left with two independent components, I_{xyz} and I_{zxy}:

$$F(\boldsymbol{I}) = 8 \left(\begin{array}{ccc|ccc|ccc} 0 & 0 & 0 & I_{xyz} & 0 & 0 & I_{xyz} & 0 & 0 \\ 0 & 0 & 0 & 0 & I_{xyz} & 0 & 0 & I_{xyz} & 0 \\ 0 & 0 & 0 & 0 & 0 & I_{zxy} & 0 & 0 & I_{zxy} \end{array}\right) \tag{7}$$

Note that equation 7 is valid only in the disordered phase, as additional components of the tensor appear in the ordered phase with the reduction of symmetry. We see that third-rank tensors do not vanish in general in the disordered phase. In addition to the E1E2 term, one can show that the third-rank tensor representing TMI and PSI scattering also contributes to forbidden reflections. A similar analysis of the fourth-rank tensor representing E2E2 processes reveals that a single independent component survives in the structure factor. With several independent components of presumably similar amplitudes contributing to the tensor structure factor in the disordered phase, one can expect a dependence of the spectrum with the chosen reflection, its azimuth, and the polarisation.

In conclusion, REXS at forbidden reflections is dominated by the E1E1 tensor in the ordered phase, but it vanishes in the disordered phase, unveiling contributions from weaker resonant processes. We expect therefore a large difference of intensity between the two phases, in favour of the ordered phase. Since different tensors are involved, we also expect a change in the energy spectrum of the forbidden reflections.

3. Experimental

A single crystal of KDP with surface normal 001 was measured at beamlines I16 of Diamond Light Source [28] and BM28 (XMaS) of the European Synchrotron Radiation Facility. The sample was enclosed in a closed-cycle cryofurnace and the temperature varied between 60 K and 400 K (I16) or 15 K and 320 K (XMaS). The lattice parameters of the orthorhombic phase are a=10.53 Å, b=10.44 Å, c=6.90 Å[29]. The 002 and 402 forbidden reflections were measured at the potassium K edge (~3.608 keV). Below the phase transition, the sample forms two types of domains related by a 90° rotation around the 001 axis. The a and b lattice parameters are not different enough to separate the 402 reflection of one domain type from the 042 reflection of the other domain type without using an analyser: the results presented below on the 402 reflection should be understood as a mixture of 402 and 042. The natural linear polarization of the beam was used. Most measurements were done without polarization analysis, with limited polarization analysis using the 111 reflection of a gold single crystal. The azimuthal reference is the (110) axis (equivalent to a (100) axis in the tetragonal basis) and the azimuth is zero when the azimuthal reference is in scattering plane. We found that KDP suffered from radiation damage from the full direct beam of I16 with its usual focus (120 μm x 30 μm, HxV). The beam was therefore unfocused and sufficiently attenuated to yield reproducible measurements during successive scans. A shift of 3 eV was found between the energy calibrations of both beamlines and we have rescaled the data presented here to the calibration of I16, which yields data more consistent with absorption spectra at the potassium K edge found in the literature [30, 31].

Figure 1 shows the temperature dependence of the 202 allowed reflection and the 002 forbidden reflection. The ferroelectric phase transition can be seen on both reflections around 120 K. The allowed reflection shows a strong broadening due to the formation of domains at low temperature. The forbidden reflection shows a huge change of intensity across the phase transition, due to the appearance (during warming) or disappearance (during cooling) of the E1E1 contribution, as explained above. The first order character of the transition is clearly seen on the integrated intensity of the 002 reflection, with a difference of critical temperature of about 2 K between warming and cooling (insert of Figure 1 bottom left). For each cycle, the integrated intensity can be reasonably fitted with a power law: $I = I_0\left(1-\frac{T}{T_c}\right)^{2\beta} + C$ (C accounting for the third rank tensors contribution, supposed temperature independent on this range), with a critical exponent $\beta \approx 0.23$. The reasonable agreement with a power law shows why this phase transition has been considered for long as second order and is now said almost second order.

Above the transition, the system displays a regular evolution. The 202 reflection sharpens but its integrated intensity remains constant. The 002 reflection also sharpens (Figure 1, bottom

Journal of Physics: Conference Series **519** (2014) 012006 doi:10.1088/1742-6596/519/1/012006

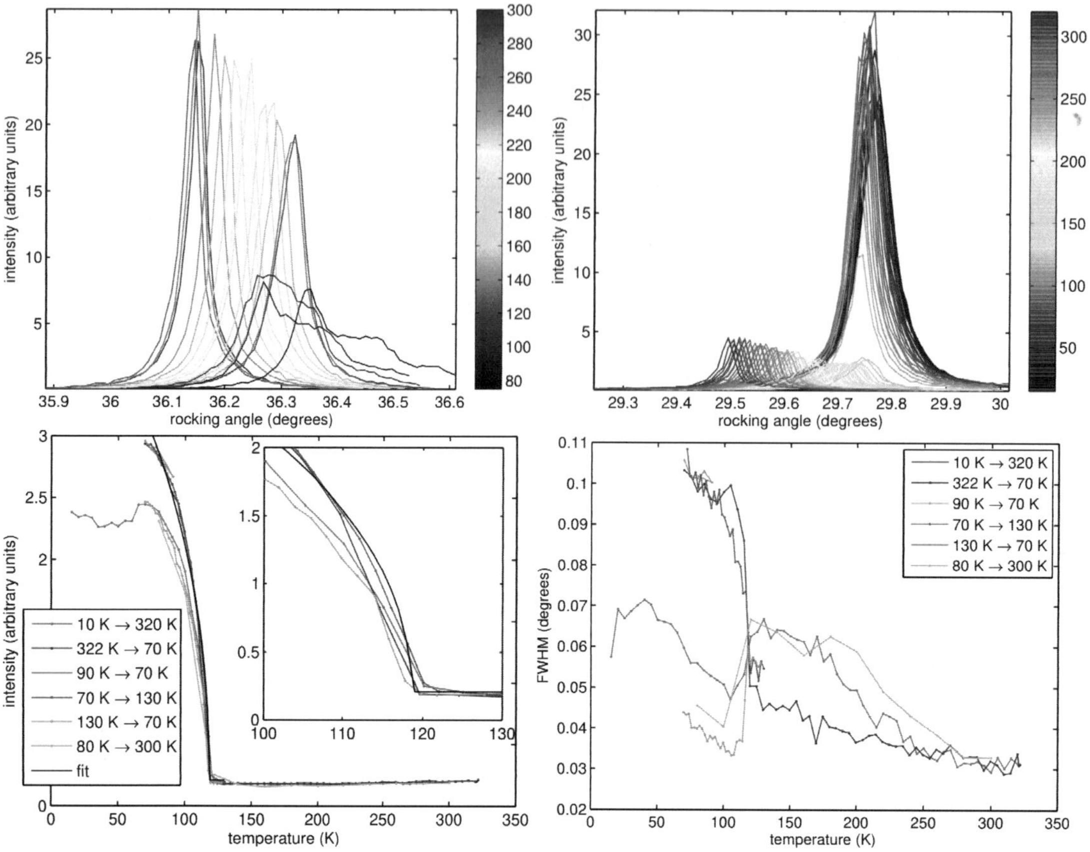

Figure 1. Temperature dependence of the 202 allowed reflection (top left) and 002 forbidden reflection (top right). The 002 reflection was measured at 3.6095 keV and azimuth -84°. Bottom: integrated intensity (left) and full-width at half-maximum (right) of the 002 reflection during several temperature cycles. A fit of the integrated intensity of the fourth temperature dependence is shown (see text for details).

right), and its integrated intensity increases smoothly with the temperature. This increase can be ascribed to TMI scattering, which grows with the activation of harmonic vibrations [18]. A reasonable fit of this increase can be obtained with the phenomenological model proposed in [32]. However PDI scattering, due to the increase of defects in the hydrogen configuration, is likely to contribute too, with a different temperature dependence. Further analysis is in progress and will be reported elsewhere [24].

The ferroelectric transition is unambiguously evidenced, not only on the rocking curve intensity at fixed energy, but also on the whole spectrum as a function of temperature (Figure 2). While there is little change of lineshape for the 002 reflection from room temperature to 122 K, the phase transition induces a strong change due to the enabling of the E1E1 channel, and remains again stable in the low temperature range (Figure 2, top row). This sudden change of intensity and lineshape was also observed on the 402 (Figure 2, middle row). In the temperature range above the phase transition, both reflections show a temperature dependence of the spectrum, which can be related to TMI and PDI scattering [24]. The effect is larger on the 402 than on the 002, largely because these effects are proportional to the square modulus of the scattering

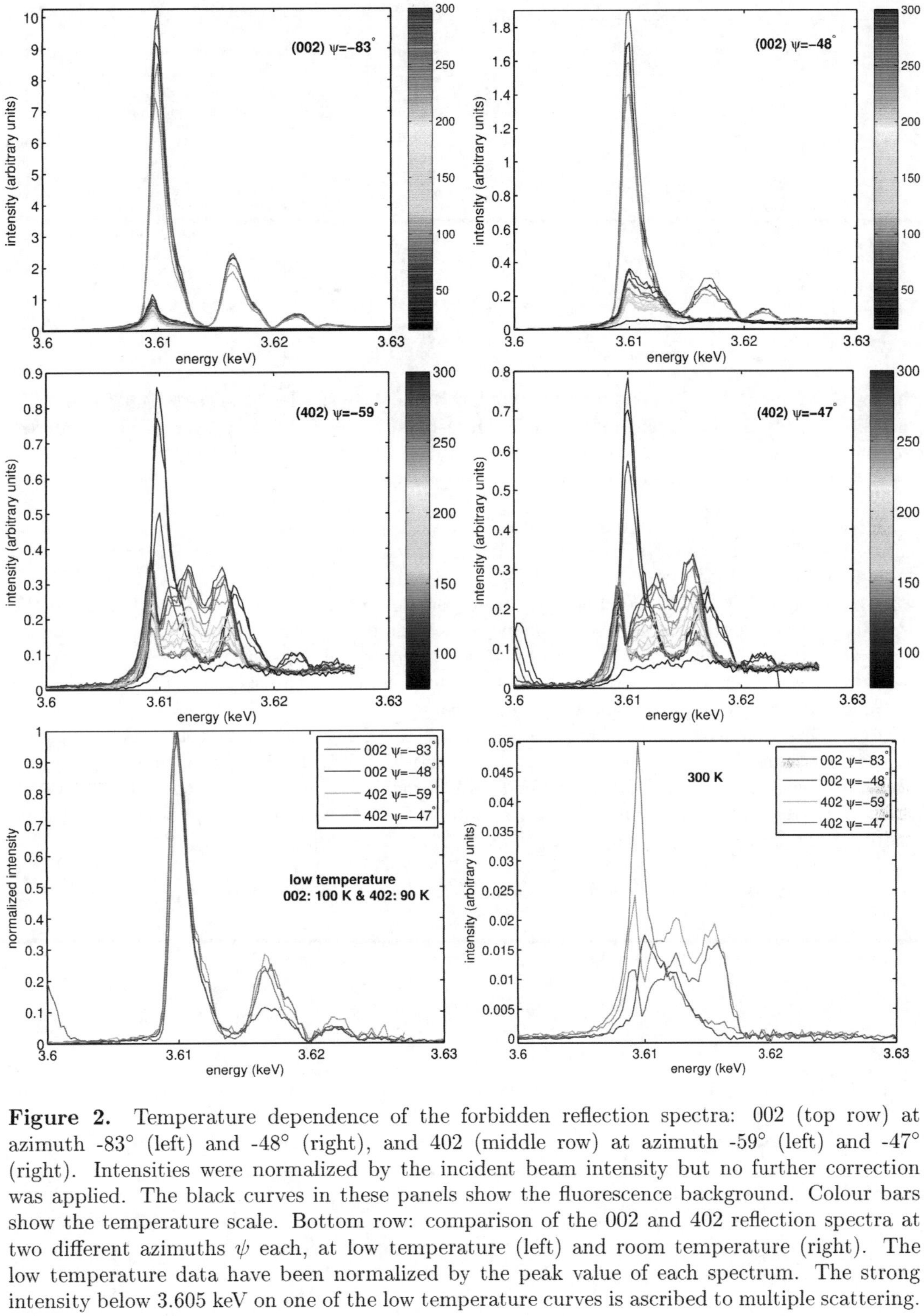

Figure 2. Temperature dependence of the forbidden reflection spectra: 002 (top row) at azimuth -83° (left) and -48° (right), and 402 (middle row) at azimuth -59° (left) and -47° (right). Intensities were normalized by the incident beam intensity but no further correction was applied. The black curves in these panels show the fluorescence background. Colour bars show the temperature scale. Bottom row: comparison of the 002 and 402 reflection spectra at two different azimuths ψ each, at low temperature (left) and room temperature (right). The low temperature data have been normalized by the peak value of each spectrum. The strong intensity below 3.605 keV on one of the low temperature curves is ascribed to multiple scattering.

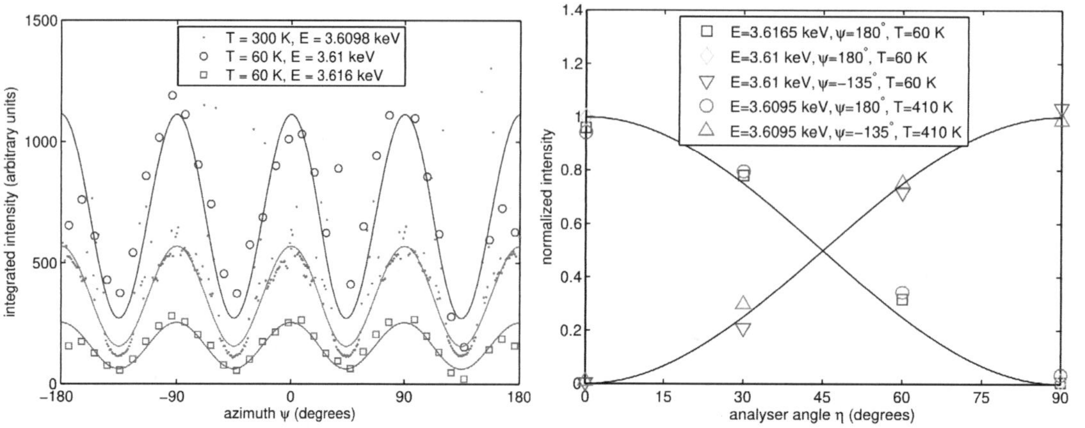

Figure 3. Azimuthal dependence without polarisation analysis (left) and polarisation analysis (right) of the 002 forbidden reflection. Intensities were integrated over rocking curves. Continuous lines are fits against the models described in the text. The continuous black lines in the right panel show $\sin^2\eta$ and $\cos^2\eta$.

vector [23]. The azimuth also plays a role. As predicted above, the low temperature spectrum is identical in shape for the 002 and 402 forbidden reflections and independent of the azimuth (Figure 2, bottom left). This demonstrates that a single tensor component (D_{xy}) dominates the intensity. There is nevertheless a small deviation for one of the curves around 3.615 keV, which we cannot explain without invoking interference with competing multiple scattering. Above the transition, the spectrum depends on the chosen reflection and on the azimuth (Figure 2, bottom right), due to the participation of several tensor components.

Let us now examine the azimuthal dependence and polarization properties of the 002 forbidden reflections (Figure 3): both phases ($T < 122$ K and $T > 122$ K) have the same polarisation properties and similar azimuthal dependence. In the following we consider only the E1E1 and E1E2 processes: the E2E2 contribution are expected to be weak compared to that of E1E2, according to calculations performed with the code FDMNES [33]. By projecting the polarisation vectors and wave vectors onto the E1E1 and E1E2 tensors given in equations 4 and 6, we obtain the structure factors for the polarisation channels $\sigma \to \sigma$ and $\sigma \to \pi$:

$$F^{\sigma\sigma}_{E1E1}(002) = -D_{xy}\cos 2\psi \tag{8}$$

$$F^{\sigma\pi}_{E1E1}(002) = D_{xy}\sin\theta\sin 2\psi \tag{9}$$

$$F^{\sigma\sigma}_{E1E2}(002) = 2I_{xyz}\sin\theta\cos 2\psi \tag{10}$$

$$F^{\sigma\pi}_{E1E2}(002) = \left[I_{zxy}\cos^2\theta + I_{xyz}\left(2 - 3\cos^2\theta\right)\right]\sin 2\psi \tag{11}$$

where θ is the Bragg angle and ψ the azimuthal angle, such that $\psi = 0$ when the 110 direction (which is the 100 direction in the tetragonal basis) lies in the scattering plane pointing away from the incident beam. For both E1E1 and E1E2, and therefore in both phases, the $\sigma \to \sigma$ channel vanishes at $\psi = 45^\circ\,[90^\circ]$ and the $\sigma \to \pi$ channel vanishes at $\psi = 0\,[90^\circ]$, as shown in the right panel of Figure 3.

In the ordered phase, we consider only the E1E1 term and neglect higher rank processes. The scattered intensity with σ-polarized incident beam and no polarization analysis is:

$$I^{\sigma}(002) \approx |F^{\sigma\sigma}_{E1E1}(002)|^2 + |F^{\sigma\pi}_{E1E1}(002)|^2 = |D_{xy}|^2\left(1 - \cos^2\theta\sin^2 2\psi\right) \tag{12}$$

The data at 60 K of Figure 3 were fitted with this equation, with $|D_{xy}|^2$ as unique free parameter. In the disordered phase, $D_{xy} = 0$ and we consider only the E1E2 contribution:

$$I^{\sigma}(002) \approx |F_{E1E2}^{\sigma\sigma}(002)|^2 + |F_{E1E2}^{\sigma\pi}(002)|^2 = I_0^{\sigma}\left(1 - a\sin^2 2\psi\right) \tag{13}$$

where I_0^{σ} and a are functions of I_{xyz}, I_{zxy} and θ, and thus depend strongly on the energy. The data at 300 K of Figure 3 were fitted with this equation, with I_0 and a as free parameters. In both phases the azimuthal dependence of the 002 reflection is four-fold and even, with maxima and minima at $\psi = 0[\pi/2]$ and $\psi = \pi/4[\pi/2]$, in agreement with experimental data. The change of symmetry of the potassium site is not visible on the azimuthal scans of the 002 reflection, which is expected since both -4 and ..2 symmetries give four-fold symmetry of the intensity along the symmetry axis. However, the contrast of the four-fold oscillation should vary strongly with the energy in the disordered phase (equation 13), while it is constant in the ordered phase (equation 12). This constant ratio allows reasonable fitting of the low temperature data, but there is unfortunately only one set of data in the disordered phase, not enough to demonstrate the change of ratio with the energy.

4. Conclusion

Forbidden reflections can be seen as a more general concept than just the cancellation of the scalar structure factor. The concept can be extended to higher rank tensors, which are relevant in REXS. In particular, while it is common to witness an order-disorder phase transition by the (dis)appearance of a non-resonant (rank 0) forbidden reflection, we have seen in this report that we can monitor the order-disorder phase transition of KDP by measuring the second rank tensor of REXS, which becomes forbidden in the disordered phase. Beyond the fundamental interest in the diffraction process for itself, this method can be useful for crystals like KDP in which the phase transition do not change the extinction conditions of non-resonant reflections. In the case of a less known structure, the disappearance of the second rank tensor during a phase transition yield a valuable information on the site symmetry.

References

[1] http://en.wikipedia.org/wiki/Monopotassium_phosphate
[2] Franken P A and Ward J F 1963 *Rev. Mod. Phys.* **35** 23–39
[3] Busch G 1938 *Helv. Phys. Acta* **11** 269
[4] Slater J C 1941 *J. Chem. Phys.* **9** 16
[5] Blinc R and Zeks B 1972 *Advances in Physics* **21** 693–757
[6] Scott J F 1974 *Rev. Mod. Phys.* **46** 83–128
[7] Nelmes R J, Meyer G M and Tibballs J E 1982 *J. Phys. C : Solid State Phys.* **15** 59–75
[8] Tibballs J E, Nelmes R J and McIntyre G J 1982 *J. Phys. C : Solid State Phys.* **15** 37–58
[9] Loiacono G, Balascio J and Osbourne W 1974 *Appl. Phys. Lett.* **24** 455
[10] van Reeuwijk S J, Puig-Molina A and Graafsma H 2001 *Phys. Rev. B* **64** 134105
[11] Lisnii B M, Levitskii R R and Baran O R 2007 *Phase Transitions* **80** 25–30
[12] Cochran W 1961 *Advances in Physics* **10** 401–420
[13] Reese W 1969 *Phys. Rev.* **181** 905
[14] Frazer B and Pepinsky R 1952 *Phys. Rev.* **85** 479–480
[15] Beale T, Beutier G, Bland S, Bombardi A, Bouchenoire L, Buñau O, Matteo S D, Fernández-Rodríguez J, Hamann-Borrero J, Herrero-Martín J, Jacques V, Johnson R, Juhin A, Matsumura T, Mazzoli C, Mulders A, Nakao H, Okamoto J, Partzsch S, Princep A, Scagnoli V, Strempfer J, Vecchini C, Wakabayashi Y, Walker H, Wermeille D, and Yamasaki Y 2012 *Eur. Phys. J. Special Topics* **208** 89–98
[16] Templeton D H and Templeton L K 1994 *Phys. Rev. B* **49** 14850
[17] Dmitrienko V E, Ovchinnikova E N, Collins S P, Nisbet G and Beutier G 2014 *Journal of Physics: Conference Series* **this volume**
[18] Dmitrienko V E and Ovchinnikova E N 2000 *Acta Cryst. A* **56** 340–347
[19] Kokubun J, Kanazava M, Ishida K and Dmitrienko V 2001 *Phys. Rev. B* **64** 073203

[20] Blume M 1994 *Magnetic Effects in Anomalous Dispersion // in Resonant Anomalous X-ray Scattering, Edited by Materlik G., Sparks C.J., Fisher K. Amsterdam* (Elsevier)
[21] Joly Y, Collins S P, Grenier S, Tolentino H C N and Santis M D 2012 *Phys. Rev. B* **86** 220101(R)
[22] Authier A 2003 *International Tables for Crystallography.* vol D
[23] Beutier G, Collins S, Nisbet G, Ovchinnikova E and Dmitrienko V 2012 *Eur. Phys. J. Special Topics* **208** 53–66
[24] in preparation
[25] Mukhamedzhanov E K, Kovalchuk M V, Borisov M M, Ovchinnikova E N, Troshkov E V and Dmitrienko V E 2010 *Crystallography Reports* **55** 174–181
[26] We spotted an error in reference [22], 1.1.4.8.5.5.: the term 333 should be absent, which is consistent with six independent components, as claimed in the reference.
[27] In the tetragonal basis with axes turned by 45°, the non-vanishing components would be the other ones: I_{xxz}, I_{xzx} and I_{zxx}.
[28] Collins S, Bombardi A, Marshall A, Williams J, Barlow G, Day A, Pearson M, Woolliscroft R, Walton R, Beutier G and Nisbet G 2010 *AIP Conf. Proc.* **1234** 303–306
[29] Levy H A, Peterson S W and Simonsen S H 1954 *Phys. Rev.* **93** 1120–1121
[30] Williams G P 2009 *X-ray data booklet* (Lawrence Berkeley National Laboratory) chap Electron binding energies
[31] Cibin G, Mottana A, Marcelli A and Brigat M F 2005 *Mineralogy and Petrology* **85** 67–87
[32] Collins S, Laundy D, Dmitrienko V, Mannix D and Thompson P 2003 *Phys. Rev. B* **68** 064110
[33] Joly Y 2001 *Phys. Rev. B* **63** 125120

Magnetic order in $GdMnO_3$ in magnetic fields

A. Skaugen[1], D.K. Shukla[1,4], R. Feyerherm[2], E. Dudzik[2], Z. Islam[3], J. Strempfer[1]

[1] Deutsches Elektronen-Synchrotron DESY, 22603 Hamburg, Germany
[2] Helmholtz-Zentrum Berlin, BESSY II, 12489 Berlin, Germany
[3] Advanced Photon Source, Argonne National Laboratory, Argonne, Illinois 60439, USA
[4] UGC DAE Consortium for Scientific Research, Khandwa Road, Indore 01, India

E-mail: arvid.skaugen@desy.de

Abstract. Resonant magnetic x-ray scattering at the Gd L_2 edge is used to investigate the magnetic order of the Gd moments in the ferroelectric phase of multiferroic $GdMnO_3$ at low temperatures and under magnetic fields. Our findings reaffirm the important role of the Gd moments in the symmetric magnetic exchange striction responsible for ferroelectricity in this compound.

1. Introduction

In recent years, strong magneto-electric coupling has been observed in orthorhombic rare-earth manganites $RMnO_3$ [1, 2, 3, 4]. Ferroelectricity in these compounds is dependent on frustration-induced cycloidal ordering of the Mn magnetic moments, which breaks spatial inversion symmetry. This is however not necessarily sufficient for creating multiferroicity, and additional possible microscopic mechanisms are still being explored [5]. The rare-earth moments have been observed to play a complex role in these materials. For example, in $DyMnO_3$ a threefold enhancement of the electric polarization is seen above 6.5 K, or when applying a magnetic field parallel to the a axis below 6.5 K, where Dy exhibits Mn-induced ordering [6, 7]. On the other hand, ferroelectricity in $GdMnO_3$ is almost entirely dependent on an applied magnetic field. The basic mechanism for ferroelectricity in $DyMnO_3$ is *antisymmetric* exchange striction between neighboring Dy and Mn moments, which displaces the rare earth ion perpendicular to the ordering wave vector [8, 9, 10]. In contrast, the *symmetric* exchange striction is hypothesized to be responsible for the polarization enhancement in $DyMnO_3$, and also for the magnetic-field induced ferroelectricity in $GdMnO_3$ [11].

In $GdMnO_3$, the Mn moments order incommensurately below $T_N^{Mn} = 43$ K [2, 12]. Below the lock-in temperature, $T' = 23$ K, the Mn order undergoes a transition to commensurate A-type AF ordering associated with weak ferromagnetism [7]. This ordering breaks spatial inversion symmetry, which in similar orthomanganites leads to ferroelectricity at the gamma point[13]. In spite of this, $GdMnO_3$ is not ferroelectric in this phase, and it is necessary to look for additional mechanisms.

Below $T_N^{Gd} = 8.2$ K both the Gd moments and the Mn moments order with propagation vector $\boldsymbol{\tau}^{Gd} = \boldsymbol{\tau}^{Mn} = \frac{1}{4}\mathbf{b}^*$. This can in theory, through the symmetric exchange interaction, cause a lattice modulation with period $q = \tau^{Gd} - \tau^{Mn} = 0$, i.e. a homogeneous lattice contraction or expansion that under certain conditions may be polar, thus leading to ferroelectricity [7].

An overview of possible modulations arising from various types of spiral magnetic ordering has been listed by Arima *et al.* [14, 15]. However, in zero field $GdMnO_3$ is only ferroelectric in a small temperature range around 5 K, even though both Mn and Gd moments are ordered with $\tau = 1/4$ down to base temperature. Ferroelectricity is induced in a larger temperature range by application of a magnetic field along the b axis. Here, one paraelectric-to-ferroelectric transition coincides with a phase boundary within the Gd ordered state [7]. We have studied the Gd ordering as a function of temperature and magnetic field in order to shed some more light on the interplay between magnetism at the Gd site and ferroelectricity in $GdMnO_3$.

2. Experiment

The experiment has been conducted at beamlines P09 at the PETRA III storage ring at Deutsches Elektronen-Synchrotron (DESY) [16] and 6-ID-B at the Advanced Photon-Source (APS) at Argonne National Laboratory. The sample was mounted and cooled in a cryomagnet with vertical field direction along the crystallographic b direction. The scattering plane was tilted slightly away from the ac plane in order to access reflections with a small k value. Unless otherwise noted, all measurements presented in this report were taken in the π-σ' polarization channel, where π (π') and σ (σ') denotes photon polarization of the incident (reflected) beam parallel and perpendicular to the scattering plane, respectively [17].

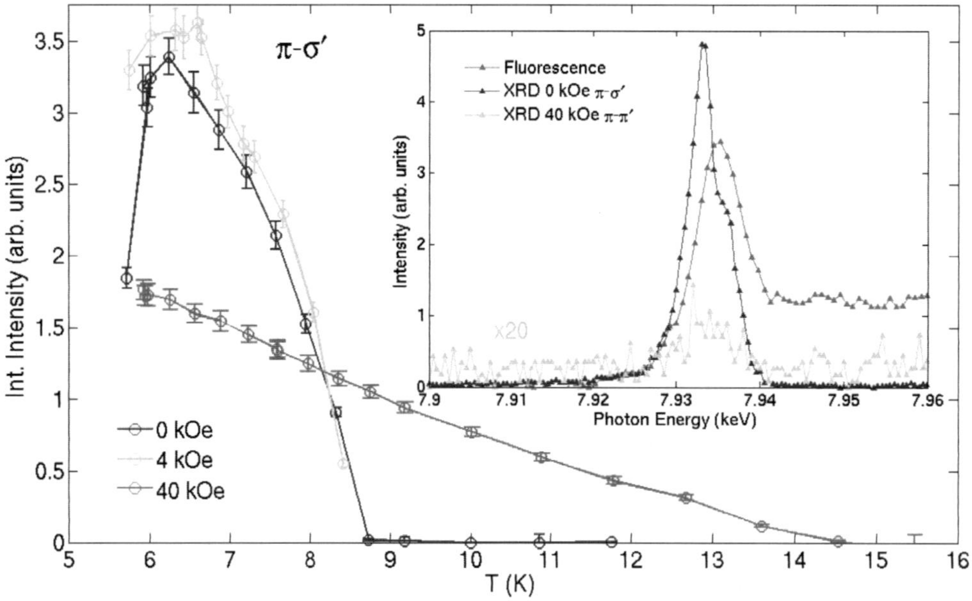

Figure 1. Temperature dependence of the (0 1/4 6) integrated intensity at different field values, measured in the π-σ' channel with a photon energy of 7.9325 keV. The inset shows the absorption corrected energy dependence across the Gd L_2 edge at T = 5.7 K, in zero field in π-σ' as well as under magnetic field in π-π', along with a fluorescence measurement of the sample.

The magnetic (0 1/4 6) reflection was investigated as function of temperature and magnetic field. At the Gd L_2 edge a strong resonant signal was measured in the π-σ' channel, confirming the magnetic origin of the reflection. Measurements were also performed with a perpendicular to the scattering plane. In this scattering geometry, at $T \sim 5.7$ K (not shown here), there is no intensity in the π-π' channel at resonance, excluding any a component of the Gd moments. This is in accordance with previous studies indicating Gd ordering predominantly along c [7].

Figure 1 shows the dependence upon temperature of the magnetic scattering intensity in the π-σ' channel with b again perpendicular to the scattering plane, for different field strengths. In zero field, an intensity enhancement around $T^* \sim 6$ K is seen. Previous measurements have shown that this is related to an increase of the b component of the Gd spins [7] and corresponds to the ferroelectric region in the low-field sector of the H-T phase-diagram from [4], ranging from about 5.5 K to 8.5 K. Application of a small field of 4 kOe results in a slight enhancement of the intensity in this region, while leaving the overall field dependence intact. Increasing the field strength to 40 kOe supresses the intensity again, but in turn extends the ordered phase to 15 K. This new transition temperature changes with field strength and does not correspond to any transition in the Mn order, indicating an onset of Gd self-ordering. With the field strength set to 40 kOe there is a weak XRD signal (which is clearly above the intensity level expected from cross-talk between the polarization channels) also in the π-π' channel, indicating a small Gd spin component also in the b direction.

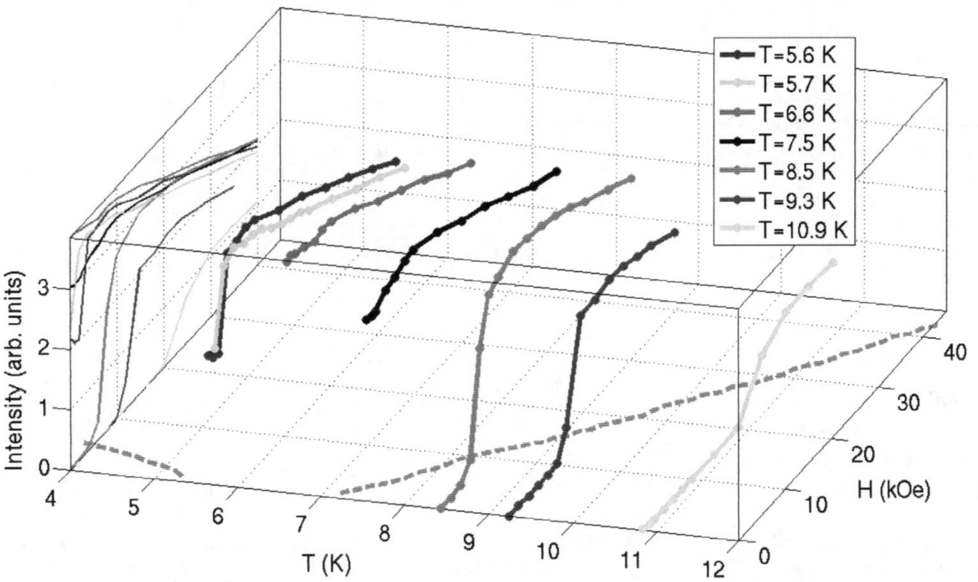

Figure 2. Field dependences of the (0 1/4 6) integrated intensity as function of **H** ∥ **b** at fixed temperatures, with projections. The orange dashed curves represent the boundaries of the ferroelectric phase from the phase diagram in [4].

In order to understand the effect of an applied magnetic field, we recorded field dependences at several different values of constant temperature. The results are shown in Figure 2. The measurements map out the H-T phase diagram of the Gd order. Comparing with the phase diagram from Ref. [4], sees that the Gd ordered phase corresponds to the ferroelectric phase in all regions.

3. Discussion

The existence of a b component of the Gd moments in the high-field region is suggestive of an important role for the symmetric magnetic exchange interaction in stabilizing the ferroelectric phase. The symmetric exchange interaction between Gd and Mn moments is proportional to $\mathbf{S}_{Gd,i} \cdot \mathbf{S}_{Mn,j}$ which is zero when the Gd moments are oriented along c, perpendicular to the Mn moments in the ab-plane. Appearance of the b component at the Gd site makes the exchange interaction non-zero, capable of causing a shift of the Gd ions towards the Mn ions. Our results therefore point to an important role of the rare earth moments for the ferroelectricity in orthomanganites, and may open up a simple route to enhance the ferroelectricity in multiferroics where the rare earth and transition metal ions order with the same wave vector.

Use of the Advanced Photon Source, an Office of Science User Facility operated for the U.S. Department of Energy (DOE) Office of Science by Argonne National Laboratory, was supported by the U.S. DOE under Contract No. DE-AC02-06CH11357.

References

[1] Fiebig M 2005 *J. Phys. D* **38** R123
[2] Kimura T, Goto T, Shintani H, Ishizaka K, Arima T and Tokura Y 2003 *Nature* **426** 55
[3] Goto T, Kimura T, Lawes G, Ramirez A P and Tokura Y 2004 *Phys. Rev. Lett.* **92** 257201
[4] Kimura T, Lawes G, Goto T, Tokura Y and Ramirez A P 2005 *Phys. Rev. B* **71** 224425
[5] Wang K F, Liu J M and Ren Z F 2009 *Advances in Physics* **58** 321–448
[6] Prokhnenko O, Feyerherm R, Dudzik E, Landsgesell S, Aliouane N, Chapon L C and Argyriou D N 2007 *Phys. Rev. Lett.* **98**(5) 057206 URL `http://link.aps.org/doi/10.1103/PhysRevLett.98.057206`
[7] Feyerherm R, Dudzik E, Wolter A U B, Valencia S, Prokhnenko O, Maljuk A, Landsgesell S, Aliouane N, Bouchenoire L, Brown S and Argyriou D N 2009 *Phys. Rev. B* **79**(13) 134426 URL `http://link.aps.org/doi/10.1103/PhysRevB.79.134426`
[8] Katsura H, Nagaosa N and Balatsky A V 2005 *Phys. Rev. Lett.* **95** 057205
[9] Sergienko I A and Dagotto E 2006 *Phys. Rev. B* **73**(9) 094434 URL `http://link.aps.org/doi/10.1103/PhysRevB.73.094434`
[10] Mostovoy M 2006 *Phys. Rev. Lett.* **96** 067601
[11] Feyerherm R, Dudzik E, Prokhnenko O and Argyriou D N 2010 *Journal of Physics: Conference Series* **200**(1) 012032
[12] Arima T, Goto T, Yamasaki Y, Miyasaka S, Ishii K, Tsubota M, Inami T, Murakami Y and Tokura Y 2005 *Phys. Rev. B* **72**(10) 100102 URL `http://link.aps.org/doi/10.1103/PhysRevB.72.100102`
[13] Aliouane N, Prokhnenko O, Feyerherm R, Mostovoy M, Strempfer J, Habicht K, Rule K C, Dudzik E, Wolter A U B, Maljuk A and Argyriou D N 2008 *J. Phys. Cond. Mat.* **20** 434215
[14] Arima T, Yamasaki Y, Goto T, Iguchi S, Ohgushi K, Miyasaka S and Tokura Y 2007 *J. Phys. Soc. Jpn.* **76** 023602
[15] Arima T 2007 *J. Phys. Soc. Jpn.* **76** 073702
[16] Strempfer J, Francoual S, Reuther D, Shukla D K, Skaugen A, Schulte-Schrepping H, Kracht T and Franz H 2013 *Journal of Synchrotron Radiation* **20** 541–549 URL `http://dx.doi.org/10.1107/S0909049513009011`
[17] Hill J P and McMorrow D F 1996 *Acta Crystallogr. A* **52** 236

Resonant x-ray scattering in $La_{1-x}Sr_{1+x}MnO_4$ ($x \geq 0.5$): Incommensurate-lattice modulation *vs.* Charge-stripe models

G Subías[1], J García[1], J Herrero-Martín[2], J Blasco[1] and M C Sánchez[1]
[1] Instituto de Ciencia de Materiales de Aragón, Departamento de Física de la Materia Condensada, CSIC-Universidad de Zaragoza, Pedro Cerbuna 12, 50009-Zaragoza, Spain
[2] ALBA Synchrotron Radiation Facility, Crta. BP 1413 Km 3.3, 08290, Cerdanyola del Vallès, Barcelona, Spain

E-mail: gloria@unizar.es

Abstract. Using resonant x-ray scattering at the Mn K-edge, we have investigated the nature of the charge and lattice modulation in the $La_{1-x}Sr_{1+x}MnO_4$ (x=0.5 and 0.6) manganites. Resonant reflections ($h\pm\varepsilon$, $h\pm\varepsilon$, 0) and ($h\pm2\varepsilon$, $h\pm2\varepsilon$, 0) of the tetragonal *I4/mmm* structure with a modulation vector of $2\varepsilon=1-x$ were found in the insulating phases of both manganites but the intensity of these reflections is much weaker for $La_{0.4}Sr_{1.6}MnO_4$. Resonant x-ray scattering data for the two samples are well explained by the presence of two types of sinusoidal modulations of the oxygen displacements, transverse and longitudinal to the tetragonal [110] direction. The amplitude of the oxygen displacements for any of the modulations decreases with the hole doping, in agreement with the change from a commensurate (x=0.5) to an incommensurate (x=0.6) ordered phase. The different polarization and azimuthal behaviour of the two sets of resonant reflections rule out any kind of stripe model composed by Mn^{3+}-like and Mn^{4+}-like charge-ordering. The maximum charge disproportionation among the different Mn atoms in the unit cell is about 0.15 e^- and 0.04 e^- for the x=0.5 and x=0.6 samples, respectively. These results thus confirm the existence of a charge-density-wave ordering in both the commensurate-phase of the half-doped $La_{0.5}Sr_{1.5}MnO_4$ and the incommensurate-phase of the over-doped $La_{0.4}Sr_{1.6}MnO_4$ manganites.

1. Introduction

In recent years, a large number of studies have been performed on the charge-orbital ordering (COO) of e_g electrons in mixed-valence manganites. Single-layered $La_{1-x}Sr_{1+x}MnO_4$ manganites have the tetragonal K_2NiF_4-type fundamental structure based on alternating stacking of MnO_2 and rock-salt type $(La,Sr)_2O_2$ layers. In this system, the concentration of e_g electrons on the Mn sites (n_e) can be described as 1-x above x=0.5 [1,2] and it exhibits the formation of superstructures that have been related to the ordering of charge and orbital of these e_g electrons at low temperatures, as well as other $RE_{1-x}(Sr/Ca)_xMnO_3$ (where RE stands for rare earth) manganites. The superlattice reflections have been observed for any $0.5 \leq x \leq 0.67$ [1,2] doping level, or commensurate and incommensurate e_g electron concentration of $0.33 \leq x \leq 0.5$. The wave vector of the structural modulation can be described by $\boldsymbol{q}=(1-x)\boldsymbol{a}^*$, where $\boldsymbol{a}^*$ is the reciprocal lattice vector. For x=0.5 (i.e. with a formal valence of $Mn^{+3.5}$), the low temperature insulating phase has been long described in terms of stripes of localized charges at atomic sites, where the $Mn^{+3.5}$ is split between Mn^{3+} ions surrounded by Jahn-

Teller oxygen distorted octahedra and Mn^{4+} ions at undistorted sites following a checkerboard pattern [3]. However, we have recently shown that the COO transition in $La_{0.5}Sr_{1.5}MnO_4$ originates from the structural transition that lowers the crystal symmetry from the tetragonal *I4/mmm* to the orthorhombic *Cmcm* driven by the condensation of three phonon modes acting on the oxygen atoms [4]. The condensation of these modes establishes the checkerboard ordering of two different Mn sites, the charge disproportionation between them being about 0.2 electrons. For x>0.5, incommensurate COO occurs and recent works have produced conflicting evidence about the nature of this incommensurate superstructure, because some studies are supporting a stripe model with charges localized at the atomic Mn sites [5] and others are indicating the formation of a charge-density wave with a periodic modulation of charge density at the atomic Mn sites [6]. The results of our resonant x-ray scattering study in $La_{0.4}Sr_{1.6}MnO_4$ [7,8] point out to the charge-density-modulation picture due to the presence of oxygen displacements sinusoidal-modulated along and perpendicular to the tetragonal [110] direction.

In the present paper, we show that the resonant x-ray scattering data at the Mn K-edge of the superlattice reflections in single-layered $La_{1-x}Sr_{1+x}MnO_4$ for all doping levels $x \geq 0.5$ are consistent with the description in terms of charge-density waves, even at commensurate doping.

2. Experimental details

Single crystals of $La_{0.5}Sr_{1.5}MnO_4$ and $La_{0.4}Sr_{1.6}MnO_4$ were grown by the floating-zone technique. The growing was carried out at 2 bars of O_2 with a growth speed ranged between 10 and 12 mm per hour [9]. X-ray diffraction measurements on crushed crystals show patterns typical of single phase tetragonal *I4/mmm* structure at room temperature. The single crystals were cut and polished in the (110) plane in the tetragonal setting.

Mn K-edge resonant x-ray scattering experiments were performed at the ID20 beam line [10] at the ESRF. The single crystals were mounted in a four-circle vertical diffractometer that is equipped with a closed-cycle helium refrigerator and a Cu (220) crystal analyser for performing σ-σ' and σ-π' polarization measurements. Energy scans across the Mn K absorption edge were measured and corrected for absorption using the experimental fluorescence. Azimuth scans at the resonance energy were recorded by rotating the sample around the scattering vector $\boldsymbol{Q}$, φ=0° for σ polarization vector parallel to the tetragonal [-110] direction. Theoretical calculations of the anomalous atomic scattering factor tensors as a function of the photon energy for the different crystallographic Mn sites were carried out using the FDMNES code [11].

3. Results and discussion

The low temperature phase in the two studied single-layered manganites is no longer tetragonal but orthorhombic, whose unit cell is metrically related to the tetragonal one as $\sqrt{2a_t} \times \sqrt{2a_t} \times c_t$ (a_t and c_t being the tetragonal lattice parameters). From here on, we refer to this small orthorhombic cell for comparison purpose. However, the superstructure cell for the $La_{0.5}Sr_{1.5}MnO_4$ sample with a commensurate doping level of x=0.5 is doubled along the orthorhombic a_o-axis (tetragonal [110]) whereas for the $La_{0.4}Sr_{1.6}MnO_4$ sample with an incommensurate concentration of x=0.6, it becomes commensurate with five cells along the a_o-axis. The formation of these superstructures is clearly confirmed by the appearance of resonant super-lattice (h±ε, h±ε, 0) and (h±2ε, h±2ε, 0) reflections in the tetragonal setting with a modulation vector 2ε=1−x. Figure 1 compares the energy dependences of some of these characteristic reflections of the superstructures of both $La_{0.5}Sr_{1.5}MnO_4$ and $La_{0.4}Sr_{1.6}MnO_4$ samples. The (1.4, 1.4, 0) and (1.6, 1.6, 0) reflections in the x=0.6 sample show the same energy and moderate azimuthal dependence as the (1.5, 1.5, 0) reflection in the x=0.5 sample, as displayed in Fig. 1(a). In the σ-σ' polarization channel, both reflections display non-resonant and resonant scattered intensities. The major difference between the two samples concerns the intensity of these superlattice reflections. The non-resonant (Thomson) scattering is about 100 times stronger for the (1.5, 1.5, 0) reflection than for the (1.4, 1.4 0) and (1.6, 1.6, 0) ones. Since the non-resonant scattering is caused by the atomic displacements breaking the tetragonal symmetry, these displacements might be significantly larger in the x=0.5 sample and decrease with the hole doping. On the other hand, the energy spectrum of both, (1.75, 1.75 0) and (1.8, 1.8, 0) reflections corresponding to the x=0.5 and x=0.6 samples, respectively, shows similar clear resonant feature in the σ-π' polarization channel, without the non-resonant signal. This indicates that these reflections are forbidden by symmetry and they shall be classified as anisotropy of the tensor of susceptibility (ATS) reflections [12]. Figure 1(b) also shows the same strong azimuthal dependence of π period for the two forbidden reflections. These ATS reflections are permitted due to the presence of a distortion of the MnO_6 octahedron. The fact that the (1.8, 1.8 0) reflection in the x=0.6 sample is reminiscent of the (1.75, 1.75 0) reflection in the half-doped sample suggests that a similar distortion of the MnO_6 octahedron occurs for both compositions. However, the intensity of the (1.75, 1.75, 0) reflection on resonance is also one order of magnitude stronger than that of the (1.8, 1.8, 0) one, which indicates that the distortion is smaller for the incommensurate structural modulation.

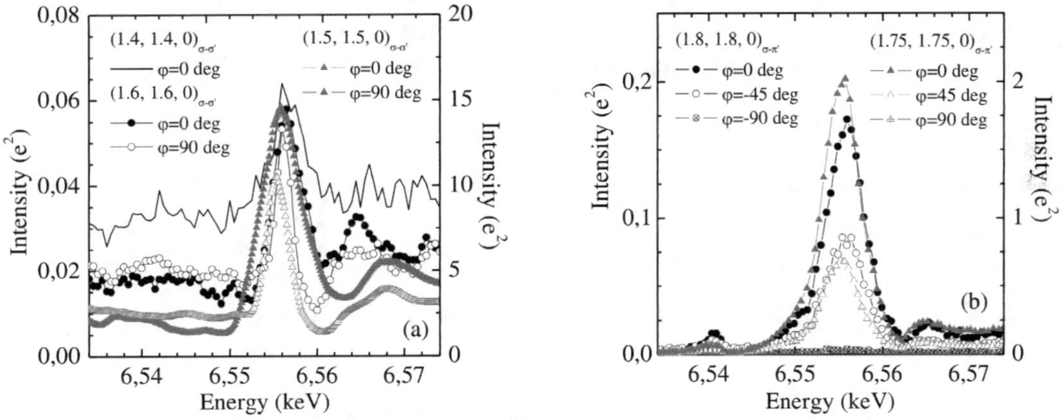

Figure 1. (a) Comparison of the energy dependence of the (1.4, 1.4, 0) and (1.6, 1.6, 0) reflections for $La_{0.4}Sr_{1.6}MnO_4$ and the (1.5, 1.5, 0) reflection for $La_{0.5}Sr_{1.5}MnO_4$ at T=80 K close to the Mn K-edge in the σ-σ' polarization channel for different azimuthal angles. (b) Comparison of the energy dependence of the (1.8, 1.8, 0) forbidden reflection for $La_{0.4}Sr_{1.6}MnO_4$ and the (1.75, 1.75, 0) forbidden reflection for $La_{0.5}Sr_{1.5}MnO_4$ at T=80 K close to the Mn K-edge in the σ-π' polarization channel for different azimuthal angles.

The resonant x-ray scattering data point to a structural modulation for the incommensurate x=0.6 sample that seems to be the same as for the commensurate x=0.5 sample, except for the magnitude of the atomic displacements. Therefore, we can consider two different models to simulate the energy, azimuthal and polarization dependence of the intensity of the reported superlattice reflections: (a) the bimodal charge-stripe model with two different Mn sites and (b) the sinusoidal-lattice modulation model with either two or five types of Mn sites for the x=0.5 and x=0.6 samples, respectively. In the case of the x=0.6 sample, the two possible bimodal charge-stripe models cannot explain the lack of σ-σ' and σ-π' intensity for the (h±0.2, h±0.2, 0) and (h±0.4, h±0.4, 0) reflections, respectively [7,8]. Thus, we propose the presence of sinusoidal-lattice modulations in the low-temperature insulating phases of both commensurate and incommensurate single-layered $La_{1-x}Sr_{1+x}MnO_4$ ($x \geq 0.5$) manganites. Since the terms in the Mn anomalous atomic scattering tensor ($f_{Mn}^{\alpha\beta}$) are mainly determined by the local structure around the Mn atom, a modulation of the oxygen atoms in the MnO_6

octahedron will induce the same kind of modulation for $f_{Mn}^{\alpha\beta}$. In the present case, a longitudinal modulation of the oxygen atoms along the a_o-axis given by $\Delta x = \Delta x_0 \cos(2\pi \times n \times 2\varepsilon)$, n being the position of the oxygen atom in the super-cell; the same kind of modulation acting on the apical oxygen atoms (along the c_o-axis) and a transverse modulation along the b_o-axis given by $\Delta y = \Delta y_0[\cos(2\pi \times n \times \varepsilon) + \sin(2\pi \times n \times \varepsilon)]$ are used. In Fig. 2, we show this superstructure model, with either two or five kinds of MnO_6 octahedra, for $La_{0.5}Sr_{1.5}MnO_4$ and $La_{0.4}Sr_{1.6}MnO_4$, respectively.

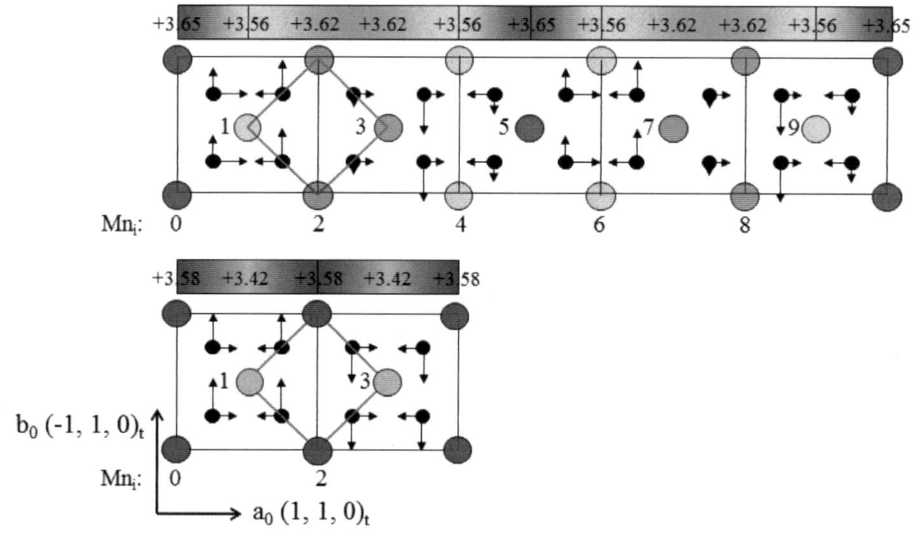

Figure 2. Proposed superstructure models for x=0.5 (lower panel) and x=0.6 (upper panel) showing longitudinal and transverse oxygen modulations in the orthorhombic a_0-b_0 plane. The respective sinusoidal Mn valence modulation as a function of position is also indicated.

We have then calculated the Mn anomalous atomic scattering tensor ($f_{Mn}^{\alpha\beta}$) for each of the non-equivalent MnO_6 octahedra using the FDMNES code [11] for a cluster radius of about 5 Å and taking into account the modulation of the MnO_6 distortion amplitude along the three orthorhombic crystallographic axes as described above. Only the diagonal f_{xx}, f_{yy} and f_{zz} components as well as the out-of-diagonal f_{xy} components of the anomalous atomic scattering tensor are found to be different from zero. The structure factors of the orthorhombic reflections (h', 0, 0) with h'=4(h±0.5) [or h'=4(h±0.25)] for x=0.5 and h'=10(h±0.4) [or h'=10(h±0.2)] for x=0.6 are given by the following expression, $F(h',0,0) = \sum_p f_{Mnp}^{\alpha\beta} \exp(i \cdot 2\pi \cdot h' \cdot x_{Mnp})$, where x_{Mnp} is the position of the Mn atom along the a_0-axis. In the model of the commensurate half-doped sample, $x_{Mnp}=0.25 \cdot p$ (p=0,1,2,3) whereas in the incommensurate x=0.6 sample, $x_{Mnp}=0.1 \cdot p$ (p=0,1,2,3,4,5,6,7,8,9). This formulation reproduces well the observed azimuthal and polarization dependencies [7, 8]. Fits to the scattered intensities allow us to obtain the maximum distortions for the longitudinal, transverse and apical modulations as well as the maximum charge disproportionation among the different Mn atoms (see figure 2). Both, oxygen displacements and charge disproportionation strongly decrease with hole doping in agreement with the experimental observation of much weaker superstructure reflections in the x=0.6 sample.

Finally, the intensity of both types of superstructure reflections decreases with increasing temperature and vanishes at T_{CO} ~ 230 K for $La_{0.5}Sr_{1.5}MnO_4$ and 240 K for $La_{0.4}Sr_{1.6}MnO_4$, as shown in Fig. 3. This indicates that the two types of reflections are correlated with the occurrence of the orthorhombic distortion. However, some differences are observed in the temperature evolution of the intensity between the super-lattice and the forbidden reflections close to T_{CO}. The slightly different temperature dependence might be related to the hierarchy of the longitudinal and transverse structural modulations driven by the phase transition.

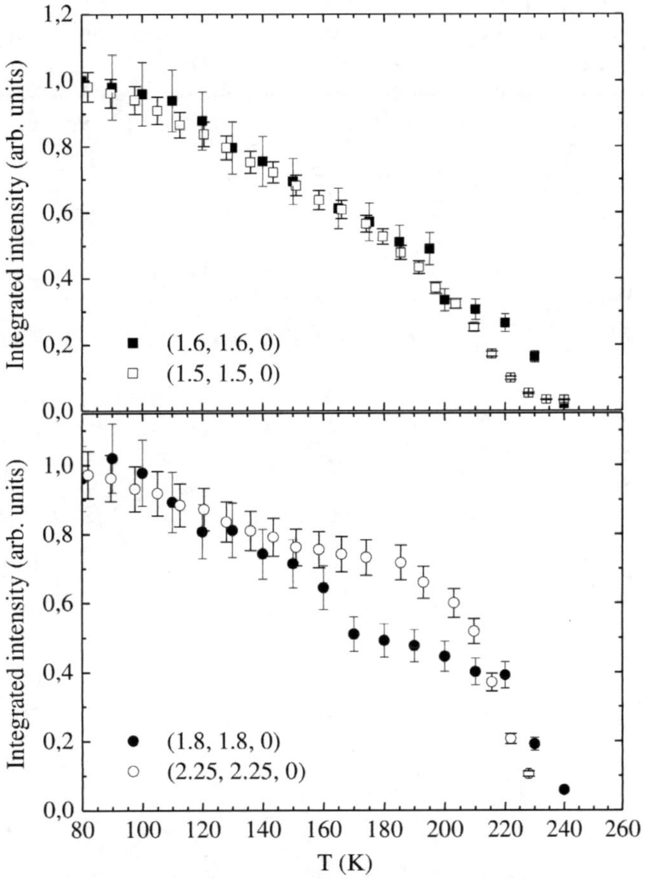

Figure 3. Integrated intensities of the (*h*±ε, *h*±ε, 0) and (*h*±2ε, *h*±2ε, 0) superstructure peaks with 2ε=1-x as a function of temperature for the two samples with doping x=0.5 (open symbols) and x=0.6 (filled symbols).

In summary, this resonant x-ray scattering study demonstrates the occurrence of (*h*±ε, *h*±ε, 0) and (*h*±2ε, *h*±2ε, 0) superstructure reflections associated to the tetragonal to orthorhombic structural transitions in the $La_{1-x}Sr_{1+x}MnO_4$ ($x \geq 0.5$) manganites with 2ε=1-x. Regardless of the commensurate (x=0.5) or incommensurate (x=0.6) character of the super-structural order, (*h*±2ε, *h*±2ε, 0) reflections originate from sinusoidal oxygen motions longitudinal to the modulation direction whereas (*h*±2ε, *h*±2ε, 0) reflections arise from sinusoidal oxygen motions transverse to the modulation direction. The different energy, polarization, azimuthal and temperature dependencies shown by the two types of superstructure reflections confirm that the 2ε modulation does not correspond to the second harmonic of the ε modulation. Therefore, the continuous charge-orbital density wave model gives a better description for the low temperature ordered phases in single-layered $La_{1-x}Sr_{1+x}MnO_4$ for all doping levels $x \geq 0.5$ than any charge-stripe model.

Journal of Physics: Conference Series **519** (2014) 012008 doi:10.1088/1742-6596/519/1/012008

References

[1] Larochelle S, Metha A, Kaneko N, Mang P K, Panchula A F, Zhou L, Arthur J and Greven M 2001 *Phys. Rev. Lett.* **87** 095502

[2] Larochelle S, Metha A, Lu L, Mang P K, Vajk O P, Kaneko N, Lynn J W, Zhou L and Greven M 2005 *Phys. Rev. B* **71** 024435

[3] Murakami Y, Kawada H, Kawata H, Tanaka M, Arima T, Moritomo Y and Tokura Y 1998 *Phys. Rev. Lett.* **80** 1932

[4] Herrero-Martín J, Blasco J, García J, Subías G and Mazzoli C 2011 *Phys.Rev. B* **83** 184101

[5] Ulbrich H, Senff D, Steffens P, Schumann O J, Sidis Y, Reutler P, Revcolevschi A and Braden M 2011 *Phys. Rev. Lett.* **106** 157201

[6] Arao M, Inoue Y, Toyoda K and Koyama Y 2011 *Phys. Rev. B* **84** 014102

[7] García J, Herrero-Martín J, Subías G, Blasco J, Andreu J S and Sánchez M C 2012 *Phys. Rev. Lett.* **109** 107202

[8] García J, Herrero-Martín J, Subías G, Blasco J and Sánchez M C 2013 *Journal of Physics: Conference Series* **430** 012107

[9] Blasco J, Sánchez M C, García J, Stankiewicz J and Herrero-Martín J 2008 *J. Cryst. Growth* **310** 3247

[10] Paolasini L, Detlefs C, M azzoli C, Wilkins S, Deen P P, Bombardi A, Kernavanois N, de Bergevin F, Yakhou F, Valade J P, Breslavetz I, Fondacaro A, Pepellin G and Bernard P 2007 *J. Synchrotron Radiat.* **14** 301

[11] Joly Y 2001 *Phys. Rev.* B **63** 125120

[12] Dmitrienko V E, Ishida K, Kirfel A and Ovchinnikova E N 2005 *Acta Crystallogr. Sect. A* **61** 481

Journal of Physics: Conference Series **519** (2014) 012009 doi:10.1088/1742-6596/519/1/012009

Resonant x-ray scattering from chiral materials: α-quartz, α-berlinite, and tellurium

Manabu Takahashi[1], **Jun-ichi Igarashi**[2]

[1] Faculty of Science and Technology, Gunma University, Kiryu, Gunma 376-8515, Japan
[2] Faculty of Science, Ibaraki University, Mito, Ibaraki 310-8512, Japan

E-mail: `mtakahas@gunma-u.ac.jp`

Abstract. We study the resonant x-ray scattering (RXS) at Si K, Al K and Te L1-edges from chiral materials, α-quartz, α-berlinite, and tellurium. For the forbidden reflections (001) and $(00\bar{1})$, we deduce the scattering matrix for the E1E1 event by summing up the local scattering matrices. The oscillation terms proportional to $\cos(3\Psi \mp \delta)$ and $\sin(3\Psi \mp \delta)$ are obtained in the spectral intensity as a function of azimuthal angle Ψ with an expression of possible phase shift δ. We evaluate the parameters which cannot be determined in the symmetry argument on the basis of underlying electronic structures given by the bond-orbital model calculation for α-quartz and α-berlinite, and by using the FLAPW band structure calculation based on the local density approximation for tellurium. The calculated spectra reproduce well the experimental results which depend on photon polarization and crystal chirality. We point out that the finite core-hole life-time and the band effect in the intermediate states may cause together the phase shift δ.

1. Introduction

Alpha-quartz (SiO_2), alpha-berlinite ($AlPO_4$), and tellurium (Te) are chiral materials, having two stereoisomeric crystal forms, the right-handed screw (space group No.152, $P3_121$) and the left-handed screw (No.154, $P3_221$). The two forms have been distinguished by using the optical activity since the discovery of Arago and Biot [1]. In order to determine the atomic positions for systems with different chirality, the anomalous x-ray scattering has been utilized [2, 3].

Recently, another method has been attempted to distinguish chirality: the resonant x-ray scattering (RXS) with circularly polarized beam [4, 5, 6], using the Si K, Al K, and Te L1-edge resonances in α-quartz, α-berlinite, and tellurium, respectively. The tensor character of scattering matrix of RXS could give rise to the intensity on the reflections forbidden in Thomson scattering [7, 8, 9]. Measuring the spectra on the forbidden reflections (001) and $(00\bar{1})$, Tanaka et al. [4, 5, 6] have found characteristic patterns depending on chirality in the spectral intensity by means of switching polarizations from the right-handed one (RCP) to the left-handed one (LCP). Regarding the oscillation of the intensity as a function of azimuthal angle, interesting phase shift has been also observed. The amount of the phase shift strongly depends on the excitation energy. It has been suggested that the phase shift might be due to the interference between the parity even E1E1 and the parity-odd E1E2 or E1M1 events [4, 5, 6]. As a possible mechanism for the phase shift, the birefringence might not be ignored. Joly et al pointed out the importance of the birefringence effect on on the analysis of the Cu L edge RXS spectra from cupric oxide [10].

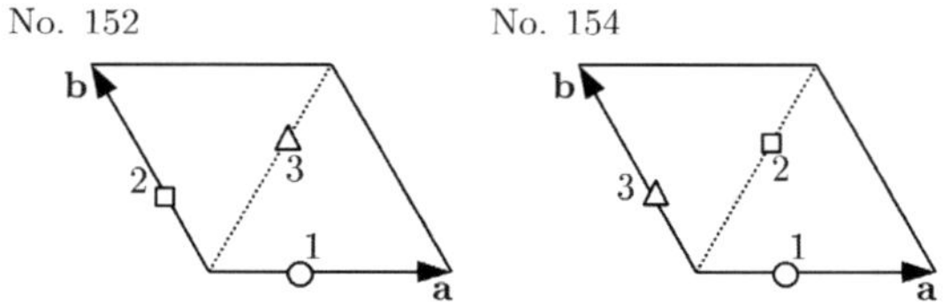

Figure 1. Si, Al, or Te atoms projected onto the a-b plane in the unit cell. Atoms labeled by 1, 2, 3 are located at $(u,0,0)$, $(0,u,\frac{1}{3})$, $(1-u,1-u,\frac{2}{3})$ for No. 152, and at $(u,0,0)$, $(1-u,1-u,\frac{1}{3})$, $(0,u,\frac{2}{3})$ for No. 154, respectively.

In this paper we discuss the dependence of the RXS intensity from chiral materials on the crystal chirality, Bragg reflection, and incident photon polarization. For the forbidden reflections(001) and $(00\bar{1})$, we deduce the general form of the scattering matrix for the E1E1 event obtained by summing up the local scattering matrices. The oscillation term proportional to $\cos(3\Psi\mp\delta)$ and $\sin(3\Psi\mp\delta)$ is obtained in the spectral intensity as a function of azimuthal angle Ψ with an expression of possible phase shift δ. We demonstrate for tellurium that the phase shift could be related to the core-hole life-time and the band effect in the intermediate states as well as the effects of the parity-odd events and the birefringence. The present paper is organized as follows. In Sec. 2, the crystal symmetry and the formula of the RXS intensity are briefly described . In Sec. 3, the calculated results are discussed. Sec. 4 is devoted to the concluding remarks.

2. Crystal Symmetry and Formula of the RXS

Bravias lattice of α-quartz, α-berlinite, and tellurium is hexagonal. As shown in Fig. 1, Si, Al, or Te atom sits at the positions $(u,0,0)$, $(0,u,\frac{1}{3})$, $(1-u,1-u,\frac{2}{3})$ for No.152, while at the positions $(u,0,0)$, $(1-u,1-u,\frac{1}{3})$, $(0,u,\frac{2}{3})$ for No.154, where $u=0.47$, 0.47, and 0.26, for Si, Al, and Te, respectively. The a-axis is a two-fold rotation axis for both No.152 and No.154, and that the crystal of No.154 is the mirror image of No.152 with respect to the a-b plane.

We construct the resonant scattering matrix based on the symmetry. We briefly summarize the derivation of formula for the RXS intensity. For the detailed derivation we refer the readers to ref. [11]. Let the incident and scattered photon polarizations be specified as x_β and x_α in Cartesian frame that the $x_1(\equiv x)$ and $x_3(\equiv z)$ axes are along the a and c axes of the hexagonal lattice. We introduce the local dipole-dipole correlation function at the core-hole site j excited by absorbing an incident photon, which is defined by

$$[\hat{\rho}_j(\epsilon)]_{\alpha,\beta}=\sum_n\langle g|\hat{x}_\alpha|n\rangle\langle n|\hat{x}_\beta|g\rangle\delta(\epsilon+\epsilon_g-\epsilon_n), \tag{1}$$

where the dipole operator $\hat{x}_\alpha$ is measured from the nuclear position of site j. Ket $|g\rangle$ represents the ground state with energy ϵ_g, and $|n\rangle$ represents the intermediate state with energy ϵ_n. Let the local dipole-dipole correlation function at site $(u,0,0)$ be $\hat{\rho}_1^{(\pm)}$ with $+$ and $-$ signs corresponding to No.152 and No.154. It takes the following matrix form according to the local symmetry:

$$\hat{\rho}_1^{(\pm)}(\epsilon)=\begin{pmatrix} a(\epsilon) & 0 & 0\\ 0 & b(\epsilon) & \pm d(\epsilon)\\ 0 & \pm d(\epsilon) & c(\epsilon)\end{pmatrix}, \tag{2}$$

where $a(\epsilon)$, $b(\epsilon)$, $c(\epsilon)$, and $d(\epsilon)$ are real functions of energy ϵ. The presence of the off-diagonal elements is due to the lack of inversion symmetry around the Si, Al, and Te atoms. The zero

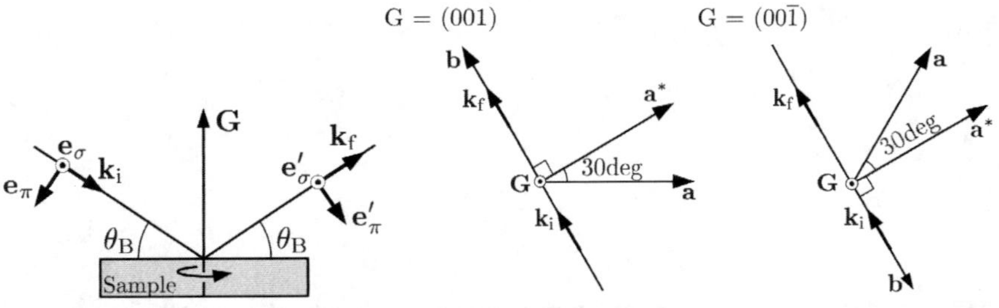

Figure 2. Scattering geometry. Left : Side view of the scattering geometry. The unit vector for the σ and π polarizations are defined as shown. Center : Top view of the scattering geometry at the azimuthal angle $\Psi = 0$ for $\mathbf{G} = (001)$, looking *down* along the c axis from the top of the axis. Right : Top view for $\mathbf{G} = (00\bar{1})$, looking *up* along the c axis from the bottom of the axis. Vectors $\mathbf{a}$ and $\mathbf{b}$ are translational vectors along the a and b axes, respectively. Vector $\mathbf{a}^*$ is the reciprocal lattice vector conjugate to $\mathbf{a}$.

components are originated from the two-fold rotation symmetry around the a-axis. The $\pm$ signs are originated from the mirror-image relation with respect to the a-b plane between No.152 and No.154. The elements $a(\epsilon)$, $b(\epsilon)$, $c(\epsilon)$, and $d(\epsilon)$ are determined on the basis of underlying electronic structure. The local correlation function $\hat{\rho}_{2(3)}^{(\pm)}(\epsilon)$ on the atomic site 2(3), are obtained by rotating $\hat{\rho}_1^{(\pm)}(\epsilon)$ by $\pm 2\pi/3(\mp 2\pi/3)$ around the c-axis. We may express the total resonant scattering matrix for scattering vector $\mathbf{G} = \mathbf{k}_\mathrm{f} - \mathbf{k}_\mathrm{i}$ as

$$\hat{M}(\mathbf{G};\omega) = \sum_j \int \frac{\hat{\rho}_j(\epsilon)\exp(-i\mathbf{G}\cdot\mathbf{r}_j)}{\omega - \epsilon + i\Gamma} \mathrm{d}\epsilon, \tag{3}$$

where $\mathbf{k}_\mathrm{i}$ and $\mathbf{k}_\mathrm{f}$ are the wave vectors for the incident and scattered x-rays, respectively. The scattering geometry is shown in fig. 2. For $\mathbf{G} = (00\bar{1})$, the total resonant scattering matrix is written as

$$\hat{M}((00\bar{1});\omega) \begin{pmatrix} A(\omega) & \mp iA(\omega) & -iB(\omega) \\ \mp iA(\omega) & -A(\omega) & \pm B(\omega) \\ -iB(\omega) & \pm B(\omega) & 0 \end{pmatrix}, \tag{4}$$

with

$$A(\omega) = \frac{3}{4}\int \frac{a(\epsilon) - b(\epsilon)}{\omega - \epsilon + i\Gamma}\mathrm{d}\epsilon, \tag{5}$$

$$B(\omega) = \frac{3}{2}\int \frac{d(\epsilon)}{\omega - \epsilon + i\Gamma}\mathrm{d}\epsilon, \tag{6}$$

where ω and Γ represent the photon energy and the life-time broadening width of the core hole state, respectively. We note that $A(\omega)$ and $B(\omega)$ are complex functions, and generally linear independent of each other. The matrix for $\mathbf{G} = (001)$ is obtained by changing the sign before the pure imaginary constant i in eq. (4).

We rotate the crystal right-handedly around $\mathbf{G}$ with azimuthal angle Ψ. Following the experimental setup by Tanaka et al., [4, 6] we define the origin of Ψ such that the scattering plane contains the b axis, as shown in Fig. 2.

In the case that the incident beam is circularly polarized and the polarization of scattered x-ray is not detected, the scattering intensity may be expressed in the following form using Stokes parameters P_1, P_2, and P_3 [12]:

$$I((00\bar{1});\omega) = I_0 - I_1\sin(3\Psi \mp \delta) - I_2\cos(3\Psi \mp \delta), \tag{7}$$

with

$$I_0 = |B|^2\cos^2\theta_{\rm B} + |A|^2\frac{1+\sin^2\theta_{\rm B}}{2} \times(1+\sin^2\theta_{\rm B} \pm 2P_2\sin\theta_{\rm B} + P_3\cos^2\theta_{\rm B}), \tag{8}$$

$$I_1 = |AB|\cos\theta_{\rm B}(P_2\cos^2\theta_{\rm B} \mp 2P_3\sin\theta_{\rm B}), \tag{9}$$

$$I_2 = \mp|AB|\cos\theta_{\rm B}(1+\sin^2\theta_{\rm B})P_1. \tag{10}$$

where the upper (lower) sign corresponds to the crystal of No.152 (No.154), and B^*A is replaced by $|AB|e^{i\delta}$. $I((001);\omega)$ is obtained by changing the sign before I_1 and I_2 in eq. (7). Right(Left) circularly polarization correspond to $P_2 = +1(-1)$. We note that the phase δ could take any value in principle. If $A(\omega)$ and $B(\omega)$ are linearly dependent on each other, δ can only take 0 or 180 deg because B^*A takes positive or negative real value. In case of $P_1 = 0$, the oscillating term as a function of azimuthal angle Ψ consist only of the term proportional to $\sin(3\Psi \mp \delta)$.

3. Calculated Results and Discussions

In order to determine the parameters $a(\epsilon)$, $b(\epsilon)$, $c(\epsilon)$, and $d(\epsilon)$, we exploit the bond-orbital model for α-quartz and α-berlinite [11], which are insulators with a wide energy gap. We make use of the FLAPW band structure calculation based on the local density approximation (LDA) for tellurium, which is a semiconductor with a tiny band gap $\sim$ 0.3eV. For α-quartz and α-berlinite, the FLAPW calculation ignoring the core-hole potential fails to reproduce the observed absorption spectral curves. On the other hand, for tellurium, the FLAPW calculation rather well reproduce the observed absorption spectral curve. Therefore, one particle approximation based on the LDA calculation ignoring the core-hole potential is expected to work safely for discussing the nature of the RXS spectra in tellurium. In the one particle approximation, the local dipole-dipole correlation function $\hat{\rho}_j(\epsilon)$ may be calculated from the single particle density matrix projected on the p-symmetric states at the scattering site. Each element in the density matrix can be explicitly expressed in terms of the FLAPW basis.

We concentrate our attention on the azimuthal angle dependence of the spectral intensity at the ω giving the main absorption peak for $\mathbf{G} = (00\bar{1})$ for α-quartz and α-berlinite belonging to No. 154. Figure 3 shows the RXS intensity as a function of azimuthal angle Ψ comparing with the observed ones in α-quartz and α-berlinite [4, 5]. Although $\mathbf{G}$ is defined by $\mathbf{k}_i - \mathbf{k}_f$ in the experiment, which is opposite to ours (see the errata in Ref. [4]), we present $\mathbf{G}$'s in our definition. The calculated intensities are smaller for RCP than those for LCP in No.154, being consistent with the experimental curves shown in panels (b) and (d), while the former is larger than the latter in No.152 (not shown). As regards the oscillation terms, we notice from the eqs. (7) and (9) that they take the form of $-a_1\sin(3\Psi - \delta)$ for RCP in No.152, and $a_1\sin(3\Psi + \delta)$ for LCP in No.154 with a_1 a positive number, and that they take the form of $b_1\sin(3\Psi - \delta)$ for LCP in No.152, and $-b_1\sin(3\Psi + \delta)$ for RCP in No.154 with b_1 a positive number. Their ratio is given by $b_1/a_1 = (|P_2|\cos^2\theta_{\rm B} + 2P_3\sin\theta_{\rm B})/(|P_2|\cos^2\theta_{\rm B} - 2P_3\sin\theta_{\rm B}) \sim 0.2$ for α-quartz and ~ 1.71 for α-berlinite, which is independent of the model. These values are quite well consistent with the observed ones. However, there are noticeable discrepancies between the calculation and the experiments. In α-quartz, the intensities for RCP in No.154 are too small in comparison with the experiment. Concerning the phase shift, the experimental curves in α-quartz seems to have the phase shift $\delta = 120 \sim 180$ deg. It is also reported that the phase shift depends on the photon energy ω in α-quartz and α-berlinite. Although the phase δ in eq. (7) could take any value in

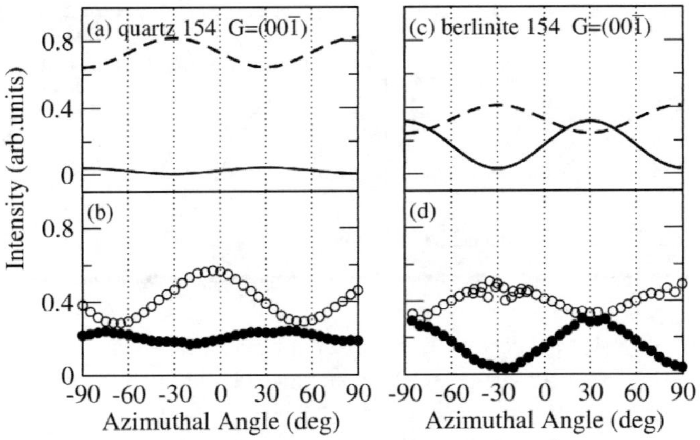

Figure 3. RXS intensity from α-quartz and α-berlinite of No.154, for $\mathbf{G} = (00\bar{1})$. The ω is fixed at the value giving the absorption peak (not shown). $P_1 = 0$, $P_3 = -0.31$, $P_2 = 0.95$ (RCP), and $P_2 = -0.95$ (LCP) and $\sin\theta_B = 0.625$ for quartz, and $P_1 = 0$, $P_3 = +0.30$, $P_2 = 0.95$ (RCP), and $P_2 = -0.95$ (LCP), and $\sin\theta_B = 0.361$ for berlinite. Lower panel shows the experimental curves reproduced from Refs. [4] and [5]. Solid curves and solid circles are for RCP, and dashed curves and open circles are for LCP.

principle, the bond-orbital model gives the phase shift $\delta = 180$ deg. independent of the photon energy. This is may caused by the fact that the four sp^3 hybrid orbitals on the scattering site are assumed not to interact with each other in our bond-orbital model, which leads to the result that $A(\omega)$ and $B(\omega)$ are linearly dependent on each other. The most serious discrepancy is that the phase shift in α-quartz looks to depend on the photon polarization and the Bragg reflection $\mathbf{G}$ [4], although the phase shift δ should be independent of them in accordance with the symmetry argument within the E1E1 approximation.

In accordance with eqs. (5) and (6), the energy dependent phase shift δ is expected to be caused by the life-time broadening Γ of the core hole and the band effect on the intermediate states. In order to elucidate these effects, we adopt the FLAPW band structure calculation for tellurium to evaluate the parameters $a(\epsilon)$, $b(\epsilon)$, $c(\epsilon)$, and $d(\epsilon)$. We assume the life-time broadening for Te 2s core hole state $\Gamma = 1.5$ eV [13]. Figure 4 shows the calculated absorption and RXS intensities as a function of photon energy ω at $\Psi = 0$, reproducing well the observed spectral shape [6]. The absorption spectra show a peak structure near the edge and broad structures in higher energy region. On the other hand, RXS spectra show almost single peak structure with small shoulder structures in higher energy region. The calculated phase shift δ is found to strongly depend on the photon energy as shown in the figure 4(center). The phase shift at the photon energy giving the RXS peak intensity is $\delta \sim 196$ deg, which is nearly equal to the observed value $\delta_{\rm exp} \sim 194$ deg. The calculated intensities are slightly smaller for RCP than for LCP in No.154 and the ratio b_1/a_1 equals to 1. These are well consistent with the observations.

4. Concluding Remarks

We have analyzed the RXS spectra at Si and Al K-edges and Te L1-edge on forbidden reflections in chiral materials, α-quartz, α-berlinite, and tellurium, using the expressions of scattering matrix for the E1E1 event. The expressions include the oscillation term as a function of azimuthal angle, which has the phase shift and the amplitude with chirality dependence.

The calculated spectra based on the bond orbital model have reproduced well main aspect of the experimental observations α-quart and α-berlinite. However, some discrepancies remain in

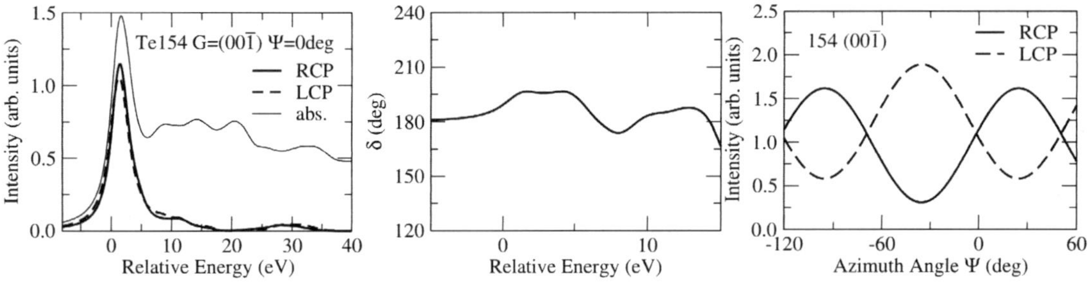

Figure 4. Calculated absorption and RXS intensities as a function of photon energy (left), phase shift δ as a function of photon energy (center), and RXS peak intensity as a function of azimuthal angle (right) in tellurium of No.154 for $\mathbf{G} = (00\bar{1})$. The energy zero corresponds to the energy difference between Te 2s level and the conduction band bottom. The life-time broadening is chosen as $\Gamma = 1.5$ eV. $P_1 = 0$, $P_3 = 0$, $P_2 = 1$ (RCP), and $P_2 = -1$ (LCP) and $\sin\theta_B = 0.212$. Solid and dashed curves are for RCP, and LCP, respectively. The azimuthal angle zero in the experiment corresponds to $\Psi = -90$ deg.

the average intensities and the phase shift of azimuthal angle oscillation in α-quartz. The phase shift depending on the photon energy has not been explained within the bond orbital model calculation. On the other hand, for tellurium the calculated spectra based on the band structure calculation have well reproduced the observations. We have demonstrated that the finite core-hole life-time and the band effect in the intermediate states can cause together the energy dependent phase shift δ in the azimuthal dependence of the scattering intensity even for the E1E1 event as well as the effects of the parity-odd events and the birefringence. It is interesting that the finite core-hole life-time causes not only the broadening of the spectra but also the phase shift. Using FLAPW band structure calculation to evaluate the dipole-dipole correlation function, and assuming the core-hole life-time broadening $\Gamma = 1.5$eV, we have successfully reproduced the absorption and RXS spectra at the Te L1 edge as functions of photon energy, and the azimuthal angle dependence of RXS intensity, including the amount of the phase shift for tellurium. For better understanding the phase shift behavior for α-quartz and α-berlinite, further studies taking account of the core hole potential and band effect are required.

Acknowledgments
This work was partially supported by a Grant-in-Aid for Scientific Research from the Ministry of Education, Culture, Sports, Science and Technology of the Japanese Government.

5. References
[1] For a historical account, see for example, S. F. Mason, *Molecular Optical Activity and the Chiral Discriminations* (Cambridge University Press, 1982).
[2] Bijvoet J M, Peerdeman A F and van Bommel J A 1951 *Nature* **168** 271
[3] de Vries A 1958 *Nature* **181** 1193
[4] Tanaka Y, Takeuchi T, Lovesey S W, Knight K S, Chainani A, Tanaka Y, Oura M, Senba Y, Ohashi H and Shin S 2012 *Phys. Rev. Lett.* **100** 145502(2008); *ibid.* **108**, 019901(E)
[5] Tanaka Y, Kojima T, Takata Y, Chainami A, Lovesey S W, Knight K S, Takeuchi T, Oura M, Senba Y, Ohashi H and Shin S 2011 *Phys. Rev. B* **81** 144104(2010); *ibid.* **84**, 219905(E)
[6] Tanaka Y, Collins S P, Lovesey S W, Matsumami M, Moriwaki M and Shin S 2012 *J. Phys.: Condens. Matter* **22** 1220(2010); *ibid.* **24**, 159905(E)
[7] Templeton D H and Templeton L K 1982 *Acta Crystallogr. A* **38** 62
[8] Dmitrienko V E 1983 *Acta Crystallogr. A* **39** 29
[9] Dmitrienko V E, Ishida K, Kirfel A and Ovchinnikova E N 2005 *Acta Crystallogr. A* **61** 481

[10] Joly Y, Collins S P, Grenier S, Tolentino H C N and Dantis M D 2012 *Phys. Rev. B* **86** 220101
[11] Igarashi J and Takahashi M 2012 *Phys. Rev. B* **86** 104116
[12] V. B berestetskii, E. M. Lifshitz, and L. P. Pitaevskii, *Quantum Electrodynamics* (Butterworth-Heinemann, 1982), Sec. 8.
[13] Krause M O and Oliver J H 1979 *J. Phys. Chem. Ref. Data* **8** 329

Coherent magnetic diffraction from the uranium M_4 edge in the multi-k magnet, USb

J A Lim[1], E Blackburn[1], G Beutier[2], F Livet[2], N Magnani[3], A Bombardi[4], R Caciuffo[3] and G H Lander[3]

[1] School of Physics and Astronomy, University of Birmingham, Birmingham, B15 2TT, United Kingdom
[2] SIMaP, CNRS Grenoble-INP - UJF, Saint-Martin d'Hères, France
[3] European Commission, Joint Research Centre, Institute for Transuranium Elements, Postfach 2340, D-76125 Karlsruhe, Germany
[4] Diamond Light Source Ltd., Rutherford Appleton Laboratory, Didcot, OX11 0QX, Oxfordshire, UK

E-mail: `joshua.ajh.lim@gmail.com`

Abstract. The slow magnetic dynamics, from seconds to kiloseconds, of the canonical 3-**k** antiferromagnet USb have been probed, using X-ray photon correlation spectroscopy (XPCS). In this work, XPCS is combined with resonant X-ray diffraction to focus on scattering at an antiferromagnetic Bragg peak. High quality coherent magnetic diffraction patterns were recorded (speckle contrast of $\sim$ 88%) and magnetic domains were observed; the number of domains increases on warming to $T^* \sim 160$ K, where the spin waves soften to zero frequency, and again on warming to $T_N = 218$ K. The intensity auto-correlation, $g_2(t)$, is primarily static over 1000 s, with a small dynamical process (change of $\sim$ 0.4%) that increases in rate close to the transitions.

1. Introduction

X-ray photon correlation spectroscopy (XPCS), also known as speckle spectroscopy, uses the correlations between coherently diffracted X-rays to access dynamics on the microsecond to kilosecond timescales, with momentum resolution [1]. To date, it has primarily been used for structural studies, but close to transitions, magnetic domain wall motion should fall into the relevant time- and length-scales. However, there are only a handful of magnetic XPCS studies, concentrating primarily on antiferromagnetic domain fluctuations [2–6] in simple antiferromagnets described by a single-**k** magnetic wavevector. In these experiments, the domain structure is encoded in the time variation of a signal closely linked to the magnetic structure. For example, coherent magnetic diffraction from a purely magnetic Bragg peak will result in interference between neighbouring domains. A high resolution coherent snapshot of this Bragg peak can then be obtained - this is known as a speckle pattern. By correlating a series of these snapshots, the characteristic timescales can be extracted. Here, we present the results of an XPCS study on the actinide multi-**k** magnet, uranium antimonide.

USb is an antiferromagnet ($T_N = 218$ K) with a 3-**k** structure, meaning that three magnetic wavevectors co-exist throughout the crystal, super-imposed on a rocksalt fcc crystal structure ($a = 6.197$ Å). Indeed, USb is the canonical example of a 3-**k** magnet, after the pioneering work of Jensen and Bak showed the spin wave spectrum could only belong to a longitudinal 3-**k** magnet [7]. The structure arises from spins pointing along local ⟨111⟩-type directions, with three equivalent wavevectors of the form [001]. This ordering acts to maintain cubic symmetry, and indeed this is confirmed by careful lattice parameter measurements [8].

The spin waves are easily observed at low temperatures, but at a temperature T^* (160 K $= T^* < T_N$) the spin wave frequency softens below the resolution limit of previous studies, and between T^* and T_N no collective excitations are observed [9, 10]. However, there is no sign of a change in the magnetic order [11, 12] and the magnetic Bragg peaks appear unaffected by this change in the dynamics [10].

Previously, it was argued [13, 14] that at T* the three independent components of the 3-**k** structure became de-phased; however, inelastic neutron scattering experiments with full polarisation analysis have shown this not to be the case [10]. In fact, the 3-**k** correlations remain even above T^* and T_N.

In a macroscopic ordered sample, it is highly likely that magnetic domains are present. Often, these arise from a need to minimise the magnetostatic energy (as is the case for ferromagnets) or due to sample topology/defects (antiferromagnets). Neighbouring domains are separated by a domain wall, which incurs some additional exchange energy cost, as well as a magnetoelastic cost due to structural differences on either side of the domain wall.

Little is known about magnetic domain dynamics in a multi-**k** system. For the case of the 3-**k** magnet, a domain wall is made of a phase slip in the magnetic propagation vector that destroys the local cubic symmetry which is necessary for, and preserved by, the 3-**k** state. For this reason, and due to exchange energy costs, the domain wall motion is expected to be relatively slow. On the other hand, in contrast to 1-**k** magnets that can couple to a lattice distortion (for example, Cr [2]), the absence of a magnetoelastic coupling in USb (i.e. pure electronic order) could result in fast domain dynamics.

The aim of this work is to use X-ray photon correlation spectroscopy to probe spatial magnetic dynamics by coherently scattering from a magnetic Bragg reflection at a resonant uranium edge from below T^* and well inside the 3-**k** state, right up to the critical regime around T_N. XPCS probes dynamics on timescales not otherwise available by other techniques, such as inelastic neutron scattering and neutron spin echo [15–17]. Previous X-ray scattering data on USb has indicated that there are multiple components to the critical scattering, with different spatial and temporal scales [18, 19].

2. Experimental Details

The key quantity obtained in an XPCS experiment is the intensity-intensity time correlation function, $g_2(\mathbf{Q}, t)$, defined as:

$$g_2(\mathbf{Q}, t) = \frac{\langle I(\mathbf{Q}, t')I(\mathbf{Q}, t' + t)\rangle}{\langle I(\mathbf{Q}, t')\rangle^2}; \quad t \geq 0 \tag{1}$$

where $I(\mathbf{Q}, t')$ is the scattering at momentum transfer vector $\mathbf{Q}$ at a time t' and the $\langle ...\rangle$ brackets denote averages over a time t'. The time t denotes the time interval of the correlation. $g_2(\mathbf{Q}, t)$ is related to the normalised intermediate scattering function, $f(\mathbf{Q}, t)$ (which contains the sample physics) and the speckle contrast, A (a beamline dependent parameter), such that:

$$g_2(\mathbf{Q}, t) = 1 + Af(\mathbf{Q}, t)^2 \tag{2}$$

In an XPCS experiment, the speckle diffraction patterns are recorded on a CCD at time instants t' (the finite counting time of exposure is ignored here, but if treated explicitly [20] only modifies the optical contrast $A \to A'$). In this discrete limit, $g_2(\mathbf{Q}, t)$ becomes:

$$g_2(\mathbf{Q}, t) = 1 + \frac{\frac{1}{N}\sum_{t'>0} I(\mathbf{Q}, t')I(\mathbf{Q}, t' + t)}{\langle I(\mathbf{Q}, t')\rangle^2} \tag{3}$$

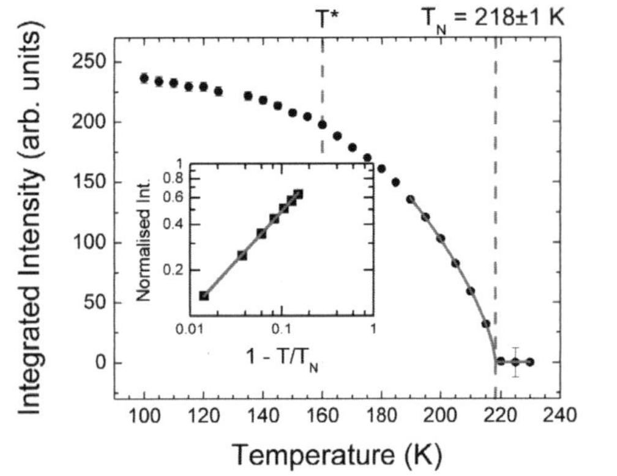

Figure 1. Integrated intensity of the **Q** = (0, 0, 3) reflection as a function of temperature, which forms a measure of the magnetic order parameter. The solid points are shown with error bars (based on the fit errors on the integrated intensity), and the solid red curve is a fit ($I = I_0\,[1 - T/T_N]^{2\beta}$ with $\beta = 0.33 \pm 0.05$). Inset: log-log plot of the **Q** = (0, 0, 3) integrated intensity against reduced temperature.

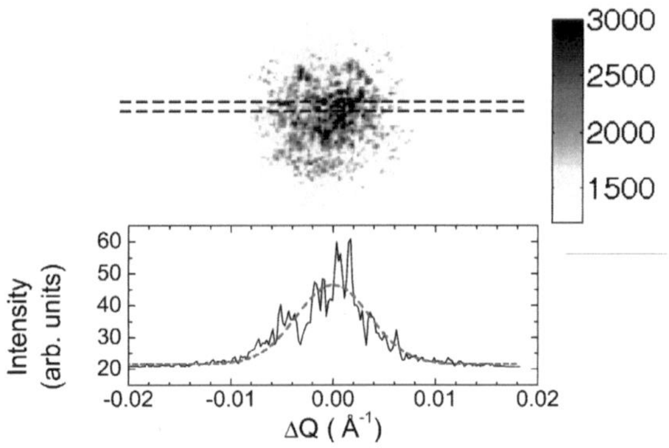

Figure 2. Upper: Typical speckle pattern obtained from the magnetic **Q** = (0, 0, 3) reflection at the M_4 resonant edge of uranium. The exposure time was 1 s and 25 x 25 μm slits were used. The black dashed lines show the region for the intensity profile in the lower panel. Lower: The intensity profile (solid line) of a slice through the speckle pattern showing intensity fluctuations that arise from the beam coherence. The red curve (Gaussian) shows the expected non-coherent intensity profile.

XPCS measurements on USb were carried out on the I16 beamline at the Diamond Light Source, UK, which was optimised for high coherence and resonant hard X-ray measurements, so is well suited for this study. The required beam coherence is generated by closing the sample slits to $\sim$25 μm^2 (this could be optimised to maximise the number of photons whilst maintaining high coherence), 70 cm before the sample, resulting in a transverse coherence length on the order of 10μm. Excellent beam stability (no decrease in $g_2(t)$ over $>$ 1000 s, see Fig. 4) and higher harmonic rejection was achieved with a set of slits 0.35 m upstream, acting as a secondary source, and additional mini-mirrors [21] (resulting in 0.1% of contamination from charge diffraction from higher harmonics at the **Q** = (0, 0, 3) position).

The key feature of the setup was an extension of the detector arm, by including a section of evacuated PVC pipe capped with Kapton® windows (detector-sample distance was 1.26 m). This was necessary to obtain high enough resolution on the detector to resolve the speckles over multiple pixels on the camera. The extension was evacuated to reduce losses and noise from air scattering. The detector used to capture snapshots of the speckle pattern was an ANDOR® iKon-M 934 (1024 x 1024 13 μm pixels), continuously cooled at -65°C.

The sample was a high quality single crystal of dimensions 2 × 2 × 2 mm, with polished faces. The average mosaic width was 0.08° at full width half maximum and the sample was mounted on a copper puck using silver paint to ensure good thermal contact and stability. Pure magnetic reflections from the uranium M_4 edge (3.728 keV) provide a large enhancement of the magnetic scattering [22], enough to mitigate against the loss of intensity from a pinhole aperture. The feasibility of this approach had been demonstrated by preliminary studies on the isostructural compound UAs at the ESRF where, in similar conditions, up to 2000 cts/s from

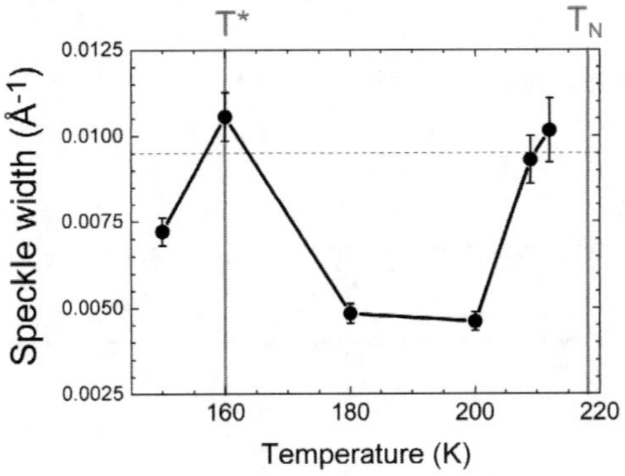

Figure 3. The extracted FWHM of the magnetic speckle pattern as a function of temperature. This is related to the average domain size and implies a decrease in the real space domain size around T^* and T_N (marked by vertical lines). The dashed horizontal line denotes the FWHM from the $\mathbf{Q}$ = (0, 0, 2) reflection, as measured at 100 K.

the $\mathbf{Q}$ = (0, 0, 1) reflection and static speckles were observable [6]. The scattering can be further enhanced by measuring at $\mathbf{Q}$ = (0, 0, 3) ($2\theta = 107.6°$), where there is a nine times intensity gain over the $\mathbf{Q}$ = (0, 0, 1) [23]. It can be seen from Eq. (15) in reference [23] that if the structure is longitudinal, as in USb, the scattering cross section is proportional to $\sin^2(\theta)$, where θ is the Bragg angle. Hence the large increase in the magnetic scattering at high Bragg angle.

The energy resolution of the setup at the M_4 edge was 4.0 ± 0.2 eV at full width at half maximum [24], corresponding to a longitudinal coherence length of $\sim 4\mu$m that is comparable to the maximum optical path difference (governed by the X-ray attenuation length and scattering geometry).

Snapshots of the speckle pattern were recorded at six different temperatures from 150 K (where we expected slow dynamics and few domains) to above both T^* and T_N (where there are expected to be many domains). Close to the transitions and in the critical regime, we expected the dynamics to slow dramatically [2–5]. Exposure times of 1, 3 and 10 seconds were used. By correlating series of images at fixed temperature, it was possible to calculate the intensity autocorrelation function to look for the characteristic timescale of the magnetic fluctuations. Speckle images were recorded 10 minutes after changing temperature; however at least 1 hour was left for the setup to stabilise, which was justified by checking that the individual values of $g_2(t)$ against acquisition time showed a random, and not a transient, response.

The calculation of $g_2(t)$ for each temperature typically used 1000 frames (where available), which corresponds to a dynamic range from ~1 second to 1 hour, although dynamics could only be reliably extracted to ~1000 s. No evidence for any ΔQ dependence to $g_2(t)$ was found, similar to other reported magnetic XPCS studies [2–5] and the whole speckle pattern was used for correlation to improve statistics. The results were independent of choosing overly large regions for analysis, due to a thresholding procedure that meant pixels with low counts did not increase the noise background [25]. A scheme using the droplet algorithm [26, 27] to count photons gave supporting results, although is not presented in this analysis.

3. Results and Discussion

The 3-**k** order parameter was measured by rocking the sample through the Bragg condition at $\mathbf{Q}$ = (0, 0, 3) position and integrating the intensity using a Voigt function (see Fig. 1). The integrated intensity near the transition was fitted with a general order parameter function

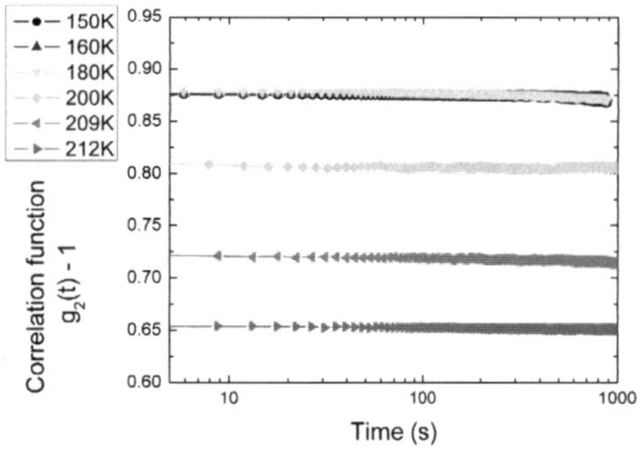

Figure 4. The calculated correlation function, $g_2(t)$, for different temperatures (error bars displayed) showing largely a static response.

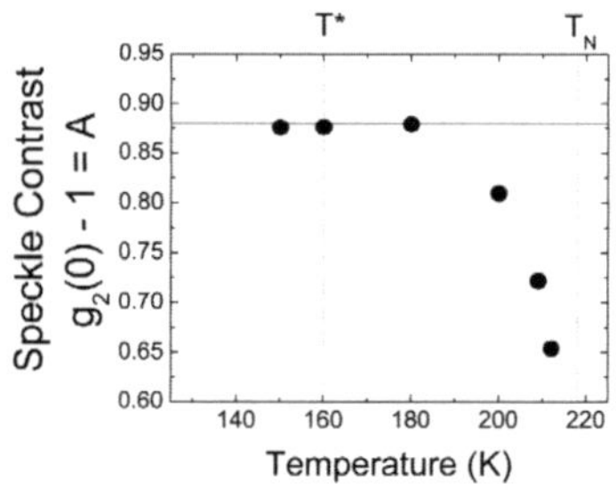

Figure 5. Magnetic speckle contrast as a function of temperature, which shows a maximum (the solid line) at $A = 0.88$.

$[I = I_0(1 - T/T_N)^{2\beta}]$ and the critical exponent, $\beta = 0.33 \pm 0.05$ and Néel temperature, $T_N = 218 \pm 1$ K were extracted, showing close agreement with published values [10, 19, 28, 29]. At the $\mathbf{Q} = (0, 0, 3)$ position for 150 K, an average of 8800 photons per second were measured.

The upper panel of figure 2 shows a typical speckle pattern, in this case measured at 150 K, that has clear and well-defined speckles. The lower panel of figure 2 shows an intensity slice through the centre of this pattern. The solid red curve (Gaussian) shows the expected diffracted intensity profile for non-coherent scattering. The large intensity fluctuations and fine structure relative to this curve indicate the high coherence and speckle contrast of the setup.

The spread of the speckle pattern on the CCD also gives information about the characteristic length scales of the magnetic domains. The central intensity profiles of the speckle patterns fit well to a Lorentzian lineshape, and the half width at half maximum, ΔQ, is partly related to the average domain size. This quantity ΔQ is a convolution of the mosaic width and the domain size. These quantities have a different Q-dependence, but with long wave-length X-rays the observations in the present experiment are too restricted to be able to separate them. The mosaic width is not expected to change with temperature, so variations in ΔQ can be interpreted as changes in the domain size.

The temperature dependence of the FWHM of the magnetic speckle pattern is shown in Figure 3. For comparison, the ΔQ spread of the structural reflections [as measured from the $\mathbf{Q} = (0, 0, 2)$ peak] is always larger than the magnetic reflections, suggesting the magnetic domain size is not limited by the structure: a feature also observed in UAs [6]. The temperature dependence shows an increase in the FWHM of the magnetic reflections close to T_N and T^* corresponding to a decrease in the magnetic domain size. Near T_N this behaviour is expected as the sample fragments into many smaller 3-**k** domains. Interestingly, this behaviour is also seen near T^* which indicates there are also changes to the domain configurations around this temperature.

From auto-correlation of the speckle patterns, $g_2(t)$ was calculated and the results (see Fig. 4) show that the correlations do not change significantly with time.

The coherence contrast can be estimated to be equal to A [recall $g_2(t) = 1 + Af(\mathbf{Q}, t)^2$ (Eqn. 2)] and at low temperatures (< 180 K) this saturates at a maximum, $A = 0.88$. The contrast was found to decrease with the scattered flux [6], but does not follow the same scaling relationship as the order parameter, indicating that other factors are also at play (see Fig. 5). High contrast for the $\mathbf{Q} = (0, 0, 2)$ charge peak was measured ($A = 0.93$) due to the increased flux and reduced 10 x 10μm slits. Nonetheless, the constant response of $g_2(t)$ reflects the excellent stability of the

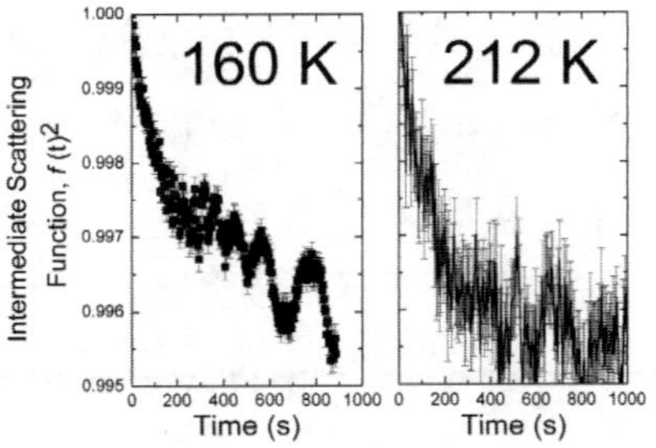

Figure 6. The normalised correlation response (the intermediate scattering function, $f(t)^2$), at temperatures near T^* and T_N zoomed in and showing the small decrease in correlation with increasing time.

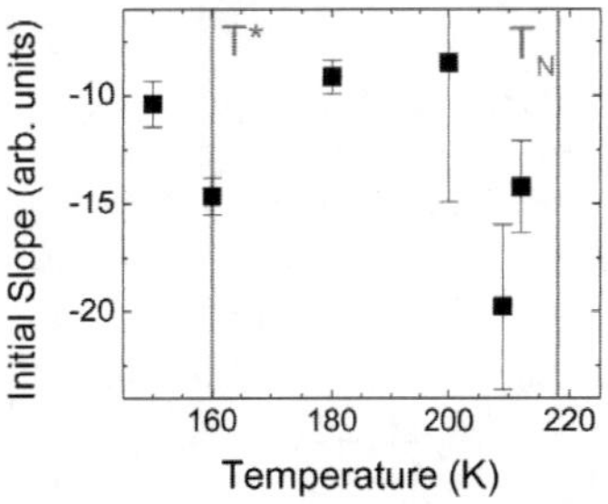

Figure 7. The initial linear gradient of the intermediate scattering function against temperature (greater negative gradient corresponds to faster dynamics) showing faster dynamics near T^* and T_N (denoted by vertical lines).

setup on I16 and the high speckle contrast shows the suitability of this beamline for correlation spectroscopy over hundreds of seconds.

This apparent time independent result corresponds with visual inspection of the speckles that appear to be static, reflecting the large static magnetisation of the 3-**k** state. As discussed earlier, the reason for the slow and quasi-static response in comparison with other 1-**k** magnetic studies may be due to the additional cubic symmetry breaking constraint of the domain wall in the 3-**k** state. Zooming in on $g_2(t)$ we see more detail and evidence for a small relaxation (see Fig. 6). At longer times, the oscillatory signal visible at 160 K is unlikely to be linked to the sample physics, but is possibly due to a property of the beamline setup.

We note that in contrast to other magnetic XPCS studies where the correlation function drops towards zero for long times [2, 5]), this is not observed in USb. We argue that this is due to static magnetic domains; however we cannot rule out small and fast fluctuations, which do not appreciably change the recorded speckle pattern or are outside of the time interval probed by this technique.

It was not possible to fit the response for all the obtained curves (nor was it always possible to clearly resolve the functional form [2, 3, 5]). A proxy for the relaxation dynamics was estimated by looking at the initial linear slope from the first 100 s (Fig. 7). This shows evidence for a faster dynamical response around T_N as expected, and also a small increase in the rate of relaxation around T^*. This dynamical picture supports the static speckle information which gives evidence for a change in the domain configuration around T_N and also T^* (see Fig. 3).

Using XPCS, we have observed a very slow response to domain motion, whilst the characteristic energies of the spin waves place the associated times in the giga-hertz regime. For this reason, it is unlikely that the changes to the magnetic domains are driving the physics at T^*, rather it appears the decrease in magnetic domain size and increase in fluctuations reflects the spin wave mode softening. It is interesting to note that this behaviour at T^*, which has been attributed to a change in itinerancy [10], can also result in changes to the magnetic domains.

4. Summary and conclusions

High quality speckle patterns were measured at the uranium M_4 resonant edge in bulk USb samples from the purely magnetic $\mathbf{Q}$ = (0, 0, 3) reflection, that showed high stability and an almost static response. The speckle contrast was found to be $\sim$ 88% in the setup using a pinhole aperture of 25 x 25 μm and demonstrates the suitability of I16 in this setup for studying magnetic domain dynamics.

There are clear change in the static and dynamical speckle patterns that show an increase in fluctuations and decrease in domain size around T_N and T^*. The cause behind the change of behaviour around T^* remains unclear, although this work shows that physics at T^* cannot only be understood by changes in itinerancy [10].

In contrast to 1-**k** magnets that show a large change in correlations and dynamics readily accessible in the sub-1000 seconds time window [2–5], in USb the magnetic domains appear static and the fluctuations slow.

These results are among the first XPCS measurements on a $5f$ electron system and this is one of the few studies to look at domains in a multi-**k** magnet. This work shows the potential of the XPCS technique which combines coherent X-rays and resonant diffraction, for study of domain dynamics that is not otherwise accessible by other means.

Acknowledgments

We thank Diamond Light Source for access to beamline I16 (MT-8384) that contributed to the results presented here.

References

[1] Grübel G and Zontone F 2004 *Journal of Alloys and Compounds* **362** 3 – 11 ISSN 0925-8388 proceedings of the Sixth International School and Symposium on Synchrotron Radiation in Natural Science (ISSRNS)
[2] Shpyrko O, Isaacs E, Logan J, Feng Y, Aeppli G, Jaramillo R, Kim H, Rosenbaum T, Zschack P, Sprung M *et al.* 2007 *Nature* **447** 68–71
[3] Seu K A, Roy S, Turner J J, Park S, Falco C M and Kevan S D 2010 *Phys. Rev. B* **82**(1) 012404
[4] Konings S, Schüßler-Langeheine C, Ott H, Weschke E, Schierle E, Zabel H and Goedkoop J B 2011 *Phys. Rev. Lett.* **106**(7) 077402
[5] Chen S W, Guo H, Seu K A, Dumesnil K, Roy S and Sinha S K 2013 *Phys. Rev. Lett.* **110**(21) 217201
[6] Yakhou F, Létoublon A, Livet F, De Boissieu M and Bley F 2001 *Journal of magnetism and magnetic materials* **233** 119–122
[7] Jensen J and Bak P 1981 *Phys. Rev. B* **23**(11) 6180–6183
[8] Knott H W, Lander G H, Mueller M H and Vogt O 1980 *Phys. Rev. B* **21**(9) 4159–4165
[9] Hagen M, Stirling W G and Lander G H 1988 *Phys. Rev. B* **37**(4) 1846–1859
[10] Lim J A, Blackburn E, Magnani N, Hiess A, Regnault L P, Caciuffo R and Lander G H 2013 *Phys. Rev. B* **87**(6) 064421
[11] Schoenes J, Frick B and Vogt O 1984 *Phys. Rev. B* **30**(11) 6578–6585
[12] Ochiai A, Suzuki Y, Shikama T, Suzuki K, Hotta E, Haga Y and Suzuki T 1994 *Physica B: Condensed Matter* **199** 616 – 618 ISSN 0921-4526
[13] Asch L, Kalvius G M, Kratzer A and Litterst F J 1994 *Hyperfine Interactions* **85**(1) 193–196 ISSN 0304-3843
[14] Lander G H and Burlet P 1995 *Physica B: Condensed Matter* **215** 7 – 21 ISSN 0921-4526
[15] Mezei F 2003 *Neutron spin echo spectroscopy* (Springer)
[16] Pappas C, Mezei F, Triolo A and Zorn R 2005 *Physica B: Condensed Matter* **356** 206 – 212 ISSN 0921-4526
[17] Squires G L 1978 *Introduction to the theory of thermal neutron scattering* (Cambridge University Press)
[18] Perry S C, Nuttall W J, Stirling W G, Lander G H and Vogt O 1996 *Phys. Rev. B* **54**(21) 15234–15237
[19] Nuttall W, Perry S, Stirling W, Mitchell P, Kilcoyne S and Cywinski R 2002 *Physica B: Condensed Matter* **315** 179 – 186 ISSN 0921-4526
[20] Jakeman E 1973 *Photon Correlation and Light Beating Spectroscopy* (Plenum, New York)
[21] Aranda M A G, Berenguer F, Bean R J, Shi X, Xiong G, Collins S P, Nave C and Robinson I K 2010 *Journal of Synchrotron Radiation* **17** 751–760
[22] Tang C C, Stirling W G, Lander G H, Gibbs D, Herzog W, Carra P, Thole B T, Mattenberger K and Vogt O 1992 *Phys. Rev. B* **46**(9) 5287–5297
[23] Hill J P and McMorrow D F 1996 *Acta Crystallographica Section A: Foundations of Crystallography* **52** 236–244
[24] Bao Z, Springell R, Walker H, Leiste H, Kuebel K, Prang R, Nisbet G, Langridge S, Ward R, Gouder T, Caciuffo R and Lander G 2013 *Physical Review B* **88** 134426
[25] Lumma D, Lurio L B, Mochrie S G J and Sutton M 2000 *Review of Scientific Instruments* **71** 3274–3289
[26] Beutier G, van der Laan G, Marty A and Livet F 2008 *The European Physical Journal Applied Physics* **42**(02) 161–167 ISSN 1286-0050

[27] Livet F, Bley F, Mainville J, Caudron R, Mochrie S, Geissler E, Dolino G, Abernathy D, Grübel G and Sutton M 2000 *Nuclear Instruments and Methods in Physics Research Section A: Accelerators, Spectrometers, Detectors and Associated Equipment* **451** 596 – 609 ISSN 0168-9002
[28] Knott H W, Lander G H, Mueller M H and Vogt O 1980 *Phys. Rev. B* **21**(9) 4159–4165
[29] Lander G H, Sinha S K, Sparlin D M and Vogt O 1978 *Phys. Rev. Lett.* **40**(8) 523–526

Journal of Physics: Conference Series **519** (2014) 012011 doi:10.1088/1742-6596/519/1/012011

Evaluation of structure models of Ho_2PdSi_3 using DAFS, *inter alia* at a satellite reflection

M Nentwich[1], M Zschornak[1], C Richter[1,2] and D C Meyer[1]

[1] Institute of Experimental Physics, TU Bergakademie Freiberg, Leipziger Straße 23, 09599 Freiberg, Germany

[2] DESY Photon Science, Notkestraße 85, 22607 Hamburg, Germany

E-mail: `Melanie.Nentwich@physik.tu-freiberg.de`

Abstract. The compounds R_2PdSi_3, with R = rare earth, exhibit a very interesting magnetic behavior with two phase transitions. Substituting one in four Si atoms by Pd in $HoSi_2$ results in a modulation of the aristotype. There are several different variants discussed in literature about the nature of the modulation of this rare-earth compound. Two of the latest models were compared: a $2 \times 2 \times 1$ layer and a $2 \times 2 \times 8$ stack. The chosen method is *Diffraction Anomalous Fine Structure* (DAFS) and was applied both experimentally and by simulation at different absorption edges and reflections, *i. a.* a satellite reflection, aiming on finding the correct crystal structure.

1. Introduction

Diffraction Anomalous Fine Structure (DAFS) is a site selective X-ray diffraction method for studying the local structure of atoms. Relying on energy scans applied at an absorption edge, the DAFS method is closely related to *X-ray Absorption Fine Structure* (XAFS) and thus offers a similar kind of spectroscopic information. As an advantage over XAFS, DAFS is sensitive to the crystallographic structure factor $|F|^2$ and, hence, it is possible to differentiate between atoms that are of the same species, but located on different Wyckoff positions, see e. g. [1].

This work focuses on the rare earth compound Ho_2PdSi_3 which is of high interest due to its complex magnetic behavior [2–6]. The structure is based on the aristotype $HoSi_2$ with the hexagonal AlB_2 structure type (space group $P6/mmm$) and the lattice parameters $a = b = 3.8148(6)$ Å and $c = 4.1073(5)$ Å [7]. The occupied Wyckoff positions are $1a$ for Ho and $2d$ for Si. The lattice parameters of the structure change slightly with the substitution of $1/4$ Si by Pd ($a = 4.076$ Å, $c = 4.021$ Å), but the AlB_2-type remains [8].

Ternary R_2TX_3 compounds with R = Sr, Y, rare earth, U, Th, T = transition metal, Al, Ga, Ge and X = Si, In, Ge, Ga, Sn have been studied by several groups in the last decades [9]. The research on this group started in 1966 [10] with the finding of the AlB_2-type structure in U_2FeSi_3 adopted from USi_2 [11]. The first work on rare earth-transition metal-silicides investigated by Raman followed in 1967 [12]. The earliest observations of a superstructure of R-T-silicides – indicated by almost doubled lattice parameters compared to their aristotypes – were reported by Chevalier *et al.* [13]. In respect to the structure of interest, in 1990, Kotsanidis *et al.* were the first to investigate Ho_2PdSi_3, finding a hexagonal cell with doubled lattice parameters a and c [2]. All following articles concerning Ho_2PdSi_3 reported structures based on the hexagonal AlB_2

Journal of Physics: Conference Series **519** (2014) 012011 doi:10.1088/1742-6596/519/1/012011

type. In contradiction to Kotsanidis *et al.* [2], Szytuła *et al.* did not find any superstructure in Ho_2PdSi_3 [3]. Several years later X-ray diffraction measurements pointed out a $2 \times 2 \times 8$ superstructure [6, 14]. Still, there remain several open questions and contradicting experimental results which are addressed in the present investigations. Tang *et al.* [14] proposed a special stacking order as shown in Fig. 1, leading to a higher symmetry than the model based on partially statistically distributed Pd atoms as proposed by Szytuła *et al.* [3]. This stacking and a $2 \times 2 \times 1$ layer (see Figure 2) originally found in Tb_2PdSi_3 [15] are considered as unit cell models within this work and compared against resonant X-ray scattering measurements.

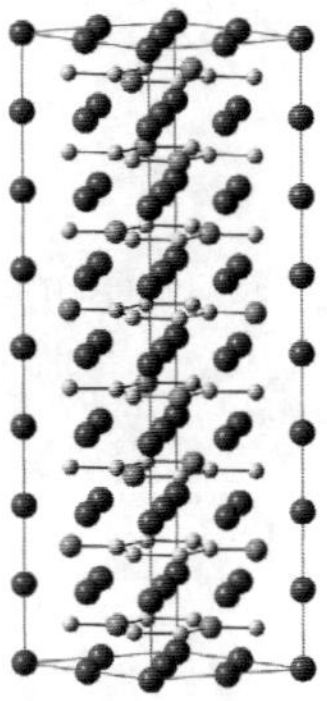

Figure 1. $2 \times 2 \times 8$ supercell D_1.

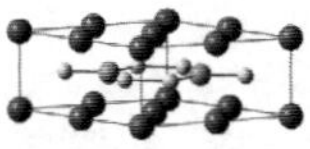

Figure 2. $2 \times 2 \times 1$ layer A.

2. Experiments

The DAFS experiments were carried out at beam lines E2 and BW1 of the former DORIS III synchrotron at DESY. For our measurements, the Bragg reflections 0 0 1, 3 0 1, 1 1 0 and $1/2\,1/2\,5/8$ were chosen carefully with respect to maximum anomalous signal at the edge, in previous considerations on the system [16]. Energy spectra of these reflections were obtained at the Pd K and Ho L_1 to Ho L_3 absorption edges evaluating the integrated intensities of rocking scans at each energy. Simultaneously, fluorescence was recorded to gain absorption spectra (XAFS). These can be used to perform an absorption correction and to find initial simulation parameters, since they contain the average spectroscopic signal of all symmetrically inequivalent resonant atoms in the structure.

The measurements at the Pd K edge at 24 347 eV were realized by using the third harmonic radiation of the Si 111 monochromator, since this energy was out of range for the fundamental. Meanwhile, the fundamental at 8 116 eV was filtered out using an aluminum absorber.

The investigated single crystals were grown by Behr *et al.* [17] and have already been studied in related work [14, 18, 19]. The chemical composition was determined as $Ho_2Pd_{1.01(1)}Si_{2.99(1)}$ corresponding to an almost ideal ratio of 2 : 1 : 3.

3. Simulation

The DAFS curves were modeled using the FDMNES code [20]. Due to the large models and hence large computing times, we started with the cluster radii `radius` and `Rself` of about 4.1 Å within the simulation. The corresponding Ho L calculations are only sensitive to the first coordination shell. However, our simulations showed that, in order to reproduce the white line in the XAFS spectrum, a radius of at least 5.4 Å must be used. Thus, each resonant Ho atom is additionally influenced by the next Ho neighbor in c-direction. This leads to the reduction of the different Pd environments to a single one and of the Ho environments to two different ones for model A and five different environments for model D_1 (determined by FDMNES). The different environments have to be averaged using the proper crystallographic weights and symmetry operations to obtain final quantities for each model, which in turn, can be compared against the measurments.

For the simulations shown below relativistic corrections as well as quadrupolar transitions have been taken into account using the refining non-standard keywords `relativism` and `quadrupole` along with the keyword `memory_save` for computational feasibility. The influence of the strongly changing absorption of the diffracted beam was taken into account using the additional option `self_abs` that performs a division by the linear absorption coefficient and, thus, corresponds to the description of integral intensities from a thick crystal in kinematic approximation.

4. Results and Discussion

The experimental data were corrected for intensity jumps due to positron injection at the synchrotron. Additionally, energy shifts of different measurements were aligned and the XAFS spectra of several reflections were averaged. For comparison, the maxima of the data – experimental as well as simulated – have been normalized to 1.

In Figure 3 the experimental and simulated XAFS data are plotted for the Ho L_3 edge. It can be seen that, in terms of position and magnitude, the resonances of the experimental XAFS fit well to the simulated curves. The whiteline of the XAFS is reproduced by both models. Model A shows a worse agreement with the experiment at energies higher than approximately 50 eV after the edge where the features are more pronounced than in the experiment. This can be related to a higher degree of order of the nearest Ho neighbors with respect to the diversity of Wyckoff sites in model D_1.

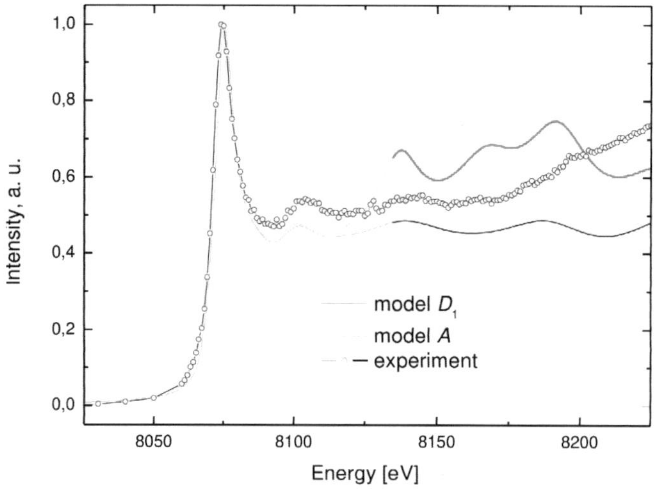

Figure 3. XAFS-Experiment (black circles) as well as simulations for model D_1 (thin blue line) and model A (thick orange line) at the Ho L_3 edge.

Figure 4 shows the DAFS-curves of the 0 0 1 reflection at the Ho L_3 edge. Both models agree well with the experiment in the edge region (edge +35 eV), beyond this region only model D_1 fits to the experiment concerning oscillation width and height.

Measured and simulated DAFS-curves of the satellite reflection $1/2\,1/2\,5/8$ at the Ho L_3 edge are compared in Figure 5. For this special reflection, only model D_1 is shown due to the expected,

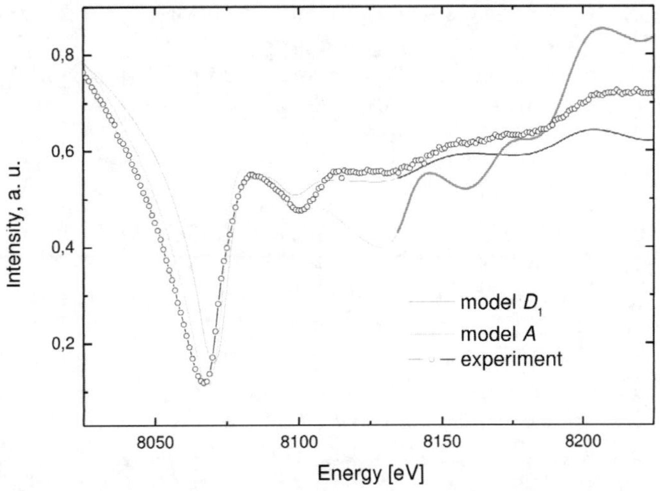

Figure 4. DAFS-Experiment (black circles) as well as simulations for model D_1 (thin blue line) and model A (thick orange line) of the $0\,0\,1$ reflection at the Ho L_3 edge.

negligible intensities for model A. In lieu of model A, we present here the double corrected intensities (I_d, bright blue) additionally to the corrected (I_c, dark blue). Figure 5 emphasizes the enormous effect of self absorption, that is underevaluated by the standard correction and overevaluated by the double correction. Both curves depict the oscillations of the experimental data. Currently a combination of integrated and maximum intensity is considered to explain the disproportionate influence of the absorption.

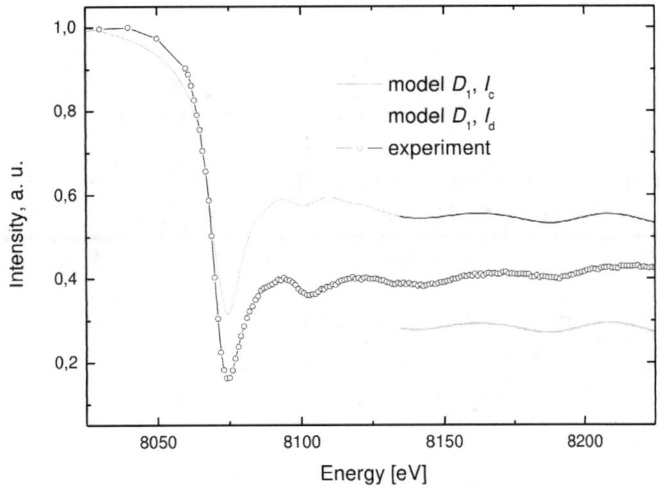

Figure 5. DAFS-Experiment (black circles) as well as simulations for corrected (I_c – thin, dark blue line) and double corrected (I_d – thick, bright blue line) model D_1 of the $^1/_2\,^1/_2\,^5/_8$ reflection at the Ho L_3 edge.

Comparing the simulated data of all reflections from both models with the according experiments, model D_1 shows a significantly better fit. Another fact supporting model D_1 is the appearance of the superlattice reflection in general.

Currently, FDMNES calculations of a further stacking model as proposed by Dshemuchadse [18] and a partially statitical distribution of Pd and Si are in progress and will be presented in future reports.

Acknowledgments

Melanie Nentwich thanks in particular the Sächsische AufbauBank for financial support via regional innovation scholarship (100111044, supported by European regional development fund, ESF). The authors in general, thank the BMBF for financial support within the project 05K10OF1, as well as the DFG within the projects ME 1433/10-1 and GE 1202/7. A part of this work was financially supported by the junior research group "PyroConvert" (100109976, supported by the European Union (European regional development fund), by the Ministry of Science and Art of Saxony (SMWK) and by GWT TUD GmbH) and the project "CryPhysConcept" (03EK3029A, supported by Federal Ministry of Education and Research (BMBF)). Additionally, we would like to thank the members of the Institute of Experimental Physics for help with words and deeds and Dr. Dmitri Novikov as our co-operation partner at DESY.

References

[1] Zschornak M, Richter C, Nentwich M, Stöcker H, Gemming S and Meyer D C 2014 *Cryst. Res. Technol.* **49**(1) 43

[2] Kotsanidis P A, Yakinthos J K and Gamari-Seale E 1990 *J. Magn. Magn. Mater.* **87**(1-2) 199

[3] Szytuła A, Hofmann M, Penc B, Ślaski M, Majumdar S, Sampathkumaran E V and Zygmunt A 1999 *J. Magn. Magn. Mater.* **202**(2-3) 365

[4] Sampathkumaran E V, Bitterlich H, Iyer K K, Löser W and Behr G 2002 *Phys. Rev. B* **66**(5) 1

[5] Frontzek M D, Kreyssig A, Doerr M, Rotter M, Behr G, Löser W, Mazilu I and Loewenhaupt M 2006 *J. Magn. Magn. Mater.* **301**(2) 398

[6] Frontzek M D, Tang F, Link P, Schneidewind A, Mignot J M, Hoffmann J U and Loewenhaupt M 2010 *J. Phys.: Conf. Ser.* **251** 012026

[7] Weitzer F, Schuster J C, Bauer J and Jounel B 1991 *J. Mater. Sci.* **26**(8) 2076

[8] Frontzek M D 2009 *PhD thesis* Technische Universität Dresden

[9] Pan Z Y, Cao C D, Bai X J, Song R B, Zheng J B and Duan L B 2013 *Chinese Phys. B* **22**(5) 056102

[10] Lourdes Pinto M d 1966 *Acta Cryst.* **21**(6) 999

[11] Brown A and Norreys J J 1961 *Nature* **191**(4783) 61

[12] Raman A and Steinfink H 1967 *Inorg. Chem.* **6**(10) 1789

[13] Chevalier B, Lejay P, Etourneau J and Hagenmuller P 1984 *Solid State Commun.* **49**(8) 753

[14] Tang F, Frontzek M D, Dshemuchadse J, Leisegang T, Zschornak M, Mietrach R, Hoffmann J U, Löser W, Gemming S, Meyer D C and Loewenhaupt M 2011 *Phys. Rev. B* **84**(10) 104105

[15] Paulose P L, Sampathkumaran E V, Bitterlich H, Behr G and Löser W 2003 *Phys. Rev. B* **21**(67) 212401

[16] Nentwich M 2012 *diploma thesis* Technische Universität Bergakademie Freiberg

[17] Behr G, Löser W, Graw G, Bitterlich H, Fink J and Schultz L 2000 *Cryst. Res. Technol.* **35**(4) 461

[18] Dshemuchadse J 2008 *diploma thesis* Technische Universität Dresden

[19] Leisegang T 2010 *PhD thesis* Technische Universität Dresden

[20] Bunău O and Joly Y 2009 *J. Phys.: Condens. Matter* **21**(34) 345501

Journal of Physics: Conference Series **519** (2014) 012012 doi:10.1088/1742-6596/519/1/012012

A new chiral phase of $BiFeO_3$ evidenced from resonant x-ray Bragg diffraction.

A Rodriguez-Fernandez[1], S W Lovesey[2, 3], S P Collins[3], G Nisbet[3] and J A Blanco[1]

[1] Physics Department, University of Oviedo, C/ Calvo Sotelo s/n Oviedo, Spain
[2] ISIS Facility, STFC Oxfordshire OX11 0QX, UK
[3] Diamond Light Source Ltd, Oxfordshire OX11 0DE, UK

E-mail: angelrf86@gmail.com

Abstract. Results that support a new chiral phase in the only multiferroic material known above room temperature have been obtained by a resonant x-ray Bragg diffraction experiment, with the incoming x-ray beam tuned near the Fe K-edge (7.1135 keV), performed in its ferroelectric phase. The R3c forbidden reflection $(0,0,9)_H$ was studied as a function of the rotation of the crystal about the Bragg wavevector in both phases, paramagnetic (700 K) and antiferromagnetic (300 K). The data gathered is consistent with a chiral structure formed by a circular cycloid propagating along $(1,1,0)_H$. Templeton and Templeton (T&T) scattering at 700 K is attributed in part to charge-like quadrupoles absent in a standard model of a cycloid in which a material vector generates all electronic states of the resonant ion. Extensive sets of azimuthal-angle data are used to infer values of three atomic multipoles in a satisfactory minimal model of the iron electronic structure, with a quadrupole (E1-E1 event) and a hexadecapole (E2-E2 event) contributing T&T scattering, plus a magnetic dipole (E1-E1).

1. Introduction

A rich variety of electronic phenomena have been revealed using scattering of x-rays at, or near, an absorption edge. These phenomena are of fundamental interest for the understanding of materials with unusual physical properties, such as $GaFeO_3$ [1], α-Fe_2O_3 [2] or the well-known Mott gap compound, V_2O_3 [3]. This technique has contributed towards the understanding of the multiferroic properties of $BiFeO_3$, the only compound known to support coupling between ferromagnetic and ferroelectric behaviours at room temperature. A better understanding of the physics that gives these material such interesting properties will help in the design of new multiferroic materials. [4]

Recently, multiferroic materials have been studied by many groups, due to the potential application for electronic devices such as sensors or multi-state memory storage-units. Of particular interest are materials that exhibit magnetoelectric properties [4-6], as is the case of the Bismuth Ferrite, $BiFeO_3$. Magnetoelectricity results from a cross-correlation of the magnetic and electric degrees of freedom of this kind of materials. This feature appears promising not only for spintronics but also for magnonics that use magnetic excitations for information processing. These materials support "spin waves" that can transport spin and hence information over macroscopic distances, without accompanying transport of charge, allowing direct control of the spin wave propagation without external magnetic fields [7].

The experimental technique used, Resonant Elastic X-ray Scattering, has advantages in the study of materials that show electronic magnetism and is able to obtain important information from the electronic properties (charge, spin and orbital degrees of freedom) of the materials, related to the angular anisotropy in the valence states, that is unobtainable by other neutron and X-ray probes [8,9]. In the case of intensities collected at space-group forbidden reflections we can access directly information about complex electronic structure which is manifest in atomic multipoles, including, magnetic charge, electric dipole, anapole, quadrupole, octupole and hexadecapole moments [2,3,10,11,12,13].

The following sections of this contribution are devoted to the discussion of the theory that is used to model a chiral long period ordering in materials belonging to the R3c space group family and presenting the Dzyaloshinky-Moriya interaction. Section 2 presents this model while Section 3 is devoted to the application of this description to a compound that displays this behaviour, $BiFeO_3$. The finally remarks are included in Section 4.

2. Theory

Bismuth Ferrite (Fig. 1) is a member of the R3c space group family, with Fe ions at 6a (Wyckoff position) sites. The symmetry related to these sites does not allow intensity by an electric dipole-electric dipole (E1-E1) event in the case of forbidden reflections of the type $(0,0,l)_H$ with l odd. However, diffraction enhanced by an electric quadrupole-electric quadrupole (E2-E2) events are allowed and can be produced by an atomic electric, time-even, hexadecapole.

Parity even multipoles are defined by the spherical tensors T^K_Q, where the positive integer K is the rank and Q the projection, with $-K \leq Q \leq K$. We represent the time average ‹...› by using angular brackets such as $\langle T^K_Q\rangle$, with a complex conjugate $\langle T^K_Q\rangle^* = (-1)^Q \langle T^K_{-Q}\rangle$. These multipoles are properties of the electronic ground-state, and the time-signature of this tensors $\langle T^K_Q\rangle$ is $(-1)^K$ [8,9]. The R3c space group symmetry allows the real part of the hexadecapole $\langle T^4_{+3}\rangle$. For the Fe ions in sites 6a in R3c structure, diffraction with wavevector $(0,0,l)_H$ is described by an electronic structure factor,

$$\Psi^K_Q = 3\left[\langle T^K_Q\rangle_1 + (-1)^l \langle T^K_Q\rangle_2\right], \tag{1}$$

where the relation between sites 1 and 2 is given by a rotation about the diagonal $\mathbf{a}_h + \mathbf{b}_h$ by 180° plus an inversion. Universal expressions for unit-cell structure factors are provided by Scagnoli and Lovesey [14]. Using (1) in an E2-E2 event, we obtain the following structure factor,

$$F_{\pi'\sigma} = \frac{3}{\sqrt{2}} \cos^3\theta \cos(3\psi) \mathrm{Re}\langle T^4_{+3}\rangle. \tag{2}$$

The contribution from the parity-odd E1-E2 events in the π'σ channel of immediate interest comes from a purely real polar quadrupole, namely, $i(3/\sqrt{5}) \cos^2\theta \langle U^2_0\rangle$ that is added to the hexadecapole contribution (2).

The presence of long range ordering due to a circular cycloid structure is responsible for a breaking of the symmetry inside the material. This motif related to the Dzyalosinky-Moriya interaction has been showed in many multiferroic materials as presented in the work by Przeniosło et al; see Model 1 in Figure 1 [15].

In the case of $BiFeO_3$, this break of symmetry could allow charge-like quadrupoles contribution (K=2) to be observed through an E1-E1 event. While electric dipole (E1-E1) transitions are normally appreciably stronger than an electric quadrupole (E2-E2) event, in this case as the diffraction from the quadrupoles is restricted due to crystal symmetry [16] and are expected to be vastly reduced in intensity.

We have defined the circular cycloid following Scagnoli and Lovesey assuming a super-cell length $L=(2n+1)a$, where a is one of the lattice parameters and n an integer related to the number of cells[14]. The integer f measures the wavevector in units of the fraction $(2p/a)(a/L)=2p/(L(2n+1))$, while the turn angle is equal to $2p/(2n+1)$. Multipoles for the super-cell are denoted by $\langle C^K_Q\rangle$. General results include: (a) Multipoles are symmetric under rotation by 180° about the axis normal to the plane of the cycloid. (b) Using the identity $C_{2x}\langle C^K_Q\rangle = (-1)^f \langle C^K_Q\rangle$ one finds the relation $\langle C^K_{-Q}\rangle = (-1)^f (-1)^K \langle C^K_Q\rangle$. (c) For given f and K all $\langle C^K_Q\rangle$ are proportional to one another. Scaling coefficients are complex and depend on both the magnitude and sign of the projection Q. (c) $\langle C^K_Q\rangle$ does not depend on n. (d) $\langle C^K_Q\rangle$ is not Hermitian. (e) $\langle C^K_Q\rangle = 0$ for $K < f$. These cycloid multipoles are not subject to the symmetry operations for sites 6a in the R3c group. For a circular cycloid rotating in the x-z plane,

$$\langle C^2_{+1}\rangle = \frac{1}{4}\left[\langle T^2_{+1} + T^2_{-1}\rangle + i\langle T^2_{+2} - T^2_{-2}\rangle\right]. \tag{3}$$

A representation of the quadrupole, $\langle T^2\rangle$, in terms of standard operators is available [17]. Note that C^2_{+1} is purely real. Magnetic diffraction by the cycloid should be created by a time-odd dipole,

$$\langle C^1_0\rangle = \frac{1}{2}\left[\langle T^1_0\rangle - i\langle T^1_{-1} - T^1_{+1}\rangle/\sqrt{2}\right]. \tag{4}$$

A dipole $\langle T^1\rangle$ is known to be a linear combination of spin, orbital moment and a dipole that expresses magnetic anisotropy [18]. However, in the absence of final state spin-orbit interactions only the orbital moment contributes at the K-edge [19].

Unit-cell structure factors in units of Ψ^2_{+1} are,

$$F_{\sigma'\sigma} = -i\,\sin(2\psi),$$

$$F_{\pi'\sigma} = t\cos\theta\,\sin\psi + u\cos^3\theta\,\cos(3\psi) - i\,\sin\theta\,\cos(2\psi)$$
$$-v(\sin\theta - i\cos\theta\,\sin\psi)\,,$$

$$F_{\sigma'\pi} = t\cos\theta\,\sin\psi + u\cos^3\theta\,\cos(3\psi) + i\,\sin\theta\,\cos(2\psi)$$
$$-v(\sin\theta + i\cos\theta\,\sin\psi)\,,$$

$$F_{\pi'\pi} = -i\,\sin^2(\theta)\sin(2\psi) - i\,v\sin(2\theta)\cos(\psi), \quad (5)$$

where θ is the Bragg angle and ψ is the azimuthal angle. Incorporating the two types of T&T mechanisms, we arrive at a unit-cell structure factor (5) that contains a second charge-like quadrupole, namely, Ψ^2_{+2}/Ψ^2_{+1} = -it, where t is a purely real value in the case of an ideal cycloid. The contribution from the hexadecapole is defined by $u=3\mathrm{Re}\langle T^4_{+3}\rangle/(\Psi^2_{+1}\sqrt{2})$. The magnetic contribution to the structure factor is $v = \Psi^1_1/(2\Psi^2_{+1})$. Since t and v both relate to an E1-E1 event they are nothing more than ratios of the appropriate multipoles that we have discussed. On the other hand, u has to include a ratio of radial integrals for E2-E2 and E1-E1 events, namely, $[q\,\{R^2\}_{sd}]^2/[\{R\}_{sp}]^2$ where $\{R\}_{sp}$ and $\{R^2\}_{sd}$.

The chiral property of the model that we present in this communication can be confirmed by a predicted contribution to intensity induced by circular polarization in the primary beam; the intensity in question is an outcome of matching chirality (helicity) in electronic structure and photons used in diffraction. The relevant intensity can be expressed in terms of unit-cell structure factors [12, 13], and after inserting expressions in (5) we arrive at,

$$\mathrm{Im}[(F_{\sigma'\pi})^* F_{\sigma'\sigma} + (F_{\pi'\pi})^* F_{\pi'\sigma}] = 2\cos\theta\,\cos\psi\,(v\sin\theta - \cos\theta\,\sin\psi)$$
$$\times\,(t\cos\theta\,\sin\psi + u\cos^3\theta\,\cos(3\psi) - v\sin\theta) \quad (6)$$

Note that the expression (6) does not vanish for v = 0, so resonant reflections have circular polarization dependence above T_N.

3. Application, $BiFeO_3$

Bismuth ferrite ($BiFeO_3$) shows both ferroic phases above room temperature. Below its Curie Temperature, $T_c \approx 1100$ K, it shows a ferroelectric behaviour with a polarization of 100 $\mu C/cm^2$. The magnetic structure of the ferric ions (Fe^{3+}) is described as a G-type antiferromagnetic ordering below the Néel temperature, $T_N = 640$ K, that coexist with a long-period incommensurate cycloid modulation ($\approx$ 620 Å) in the hexagonal plane. The propagation vector of the magnetic order is very small (k~0) and therefore it appears very close to structural Bragg reflections, as shown in Figure 1 [21,22]. The spiral propagates along the vector $(1,1,0)_H$ with the dipole moments rotate in the $(-1,1,0)_H$ plane, defined by the spontaneous electric polarization along $[0,0,1]_H$ and the propagation vector.

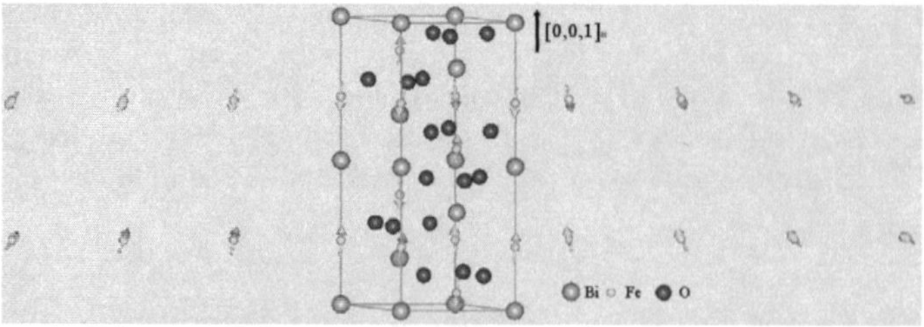

Figure 1. Scheme of the crystal and magnetic structures of bismuth ferrite ($BiFeO_3$), hexagonal $[0,0,1]_H$ axis is in the vertical. The arrows indicate the directions of Fe magnetic dipoles at room temperature.

The resonant x-ray experiments were performed at the Diamond Light Source (UK), with the synchrotron working at the energy of 3 GeV in the "top-up" mode. The beamline used was I16, equipped with a 6-circle kappa diffractometer and a focusing optical system that gives a photon flux on the sample position of 10^{13} photon/s and a beam size of 180×20 μm^2. The horizontally polarized beam, σ, delivered by a linear undulator was tuned near the Fe K-edge (7.1135 keV). As the high-quality single crystal used for the experiment (sizes 5 x 5mm^2 and a thickness of 0.5 mm) had a face perpendicular to the $(0,0,1)_H$ direction the experiment was performed with a specular geometry. While it is possible that the sample supports different domains, the very small size of the primary beam gives some confidence that a single domain is illuminated in our experiments. Indeed, we go on to show that, a chiral motif and a single domain are implied by our findings for the magnetically-ordered state.

The reflection studied was the forbidden $(0,0,9)_H$ in the space group #161. As is possible to observe from Figure 3 in ref (23), the weak contribution to the intensity by Bragg diffraction due to angular anisotropy is strongly enhanced by an atomic resonance. Two different studies were performed for the $(0,0,9)_H$ reflection at two different temperatures, below 300 K and above 700 K (the Néel Temperature). Following the scheme from Figure 2, all data collected were obtained in the rotated channel of polarization π'σ.

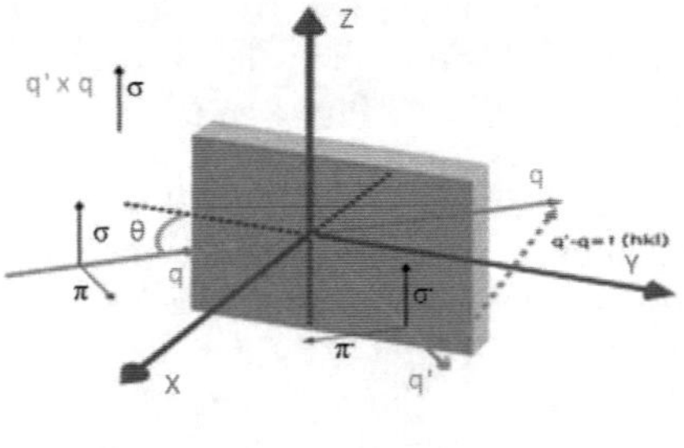

Figure 2. Scheme of the Cartesian coordinates (x, y, z) and x-ray polarization wavevectors. The plane of scattering defined by primary (**q**) and secondary (**q'**) wavevectors is parallel to the x-y plane. The components of the polarization labelled σ and σ' is normal to the plane and parallel to the z-axis, and polarization labelled π and π' lies in the plane of scattering.

The azimuthal scans presented in Figure 3 were obtained by performing "θ scans" with the detector around the Bragg condition for different azimuthal angles. This is one of the most common methods for performing azimuthal scans together with direct "ψ scans". Alternatively, one can measure the integrated intensity from a crystal analyzer, which may be pursued in the future. Coming back to the experimental method used in this experiment, the values displayed in Figure 3 are the integrated intensity of each of the curves normalized to the intensity of the primary x-ray beam. Due to the small penetration depth of an allowed Bragg spot $(0,0,6)_H$, where the diffraction strongly affacted by dynamical processes, we have not used this kind of reflection to normalize a forbidden Bragg spot $(0,0,9)_H$ that, due to its weakness has a kinematical behaviour, a larger penetration depth and is less affected by possible defects from the surface.

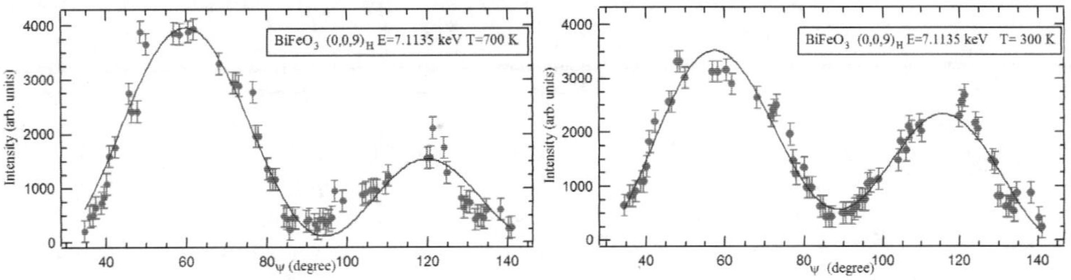

Figure 3. Templeton-Templeton scattering, at a reflection $(0,0,9)_H$ forbidden in R3c, as a function of azimuthal angle (red dots). (a) Shows data collected at the paramagnetic phase (700 K) while (b) present that of the antiferromagnetic phase (300 K). Solid line are a fit to our model of diffraction by a motif of charge-like quadrupoles and hexadecapoles, namely, $|F_{\pi'\sigma}|^2$ and expression (5) with v = 0 (v ≠ 0) for the paramagnetic (antiferromagnetic) case. Charge-like multipoles, t and u, are given the same values for both phases. Data taken from A. Rodriguez-Fernandez et al. [23]

One of the more complex steps in such kind of experiments is the extraction of the data coming directly from the intensity related to the angular anisotropy. Multiple scattering peaks, also known as Renninger reflections, were discriminated using a Matlab program available at the Beamline I16, together with the selection of optimum positions by performing azimuthal scans in the case of the data obtained at room temperature, defining flat positions between peaks and avoiding the Renninger effect (therefore measured points in the azimuth dependence are not equidistant). Due to the fact that we have collected resonant x-ray data for a certain selected reflection at different parts of the single crystal, we consider that the experimental data shown in Figure 3 are related to the resonant event rather than to the tail of the Renninger effect.

Table 1. Values of the parameters used for the fitting for the two temperatures studied. In the case of 700K the parameter v was set to 0, see text for more details.

Parameter	700K	300K
t	1.19 ± 0.07	1.19 ± 0.07
u	$- 6.20 \pm 0.16$	$- 6.20 \pm 0.16$
v	-	0.673 ± 0.014

The Templeton-Templeton (T&T) scattering reported in Figure 3 taken from A Rodriguez et al. [23] shows the dependence at the two temperatures under study (3a) 700 K and (3b) 300 K ($T_N \approx 640$ K), both curves were obtained in the polarization channel $\pi'\sigma$ (Figure 2). From the data presented it can be observed that a possible E2-E2 event together with a contribution from the parity-odd E1-E2 resonance cannot explain the azimuthal dependence, these two contributions are in phase quadrature and they cannot interfere to break the pure six-fold periodicity that is lacking in Figure 3a (700 K).

We have simulated the Templeton-Templeton scattering using circular polarized light based on the values of refined multipoles from Table 1 obtained from the fitting to the experimental data (i.e., the values of all the multipoles therein, t, u and v). Figure 4 represents the different expected signal for scattering collected above and below the Néel temperature, $T_N \approx 640$ K. The theoretical dependences are adequately calculated by a model with minimal complexity. It includes Templeton and Templeton scattering [20] from charge-like quadrupoles and hexadecapoles. The contribution from quadrupoles is allowed because they participate in a chiral, circular cycloid, otherwise their contribution is forbidden in R3c. The chiral property of our model would be confirmed by a predicted contribution to intensity induced by circular polarization in the primary beam; the intensity in question is an outcome of matching chirality (helicity) in electronic structure and photons used in diffraction.

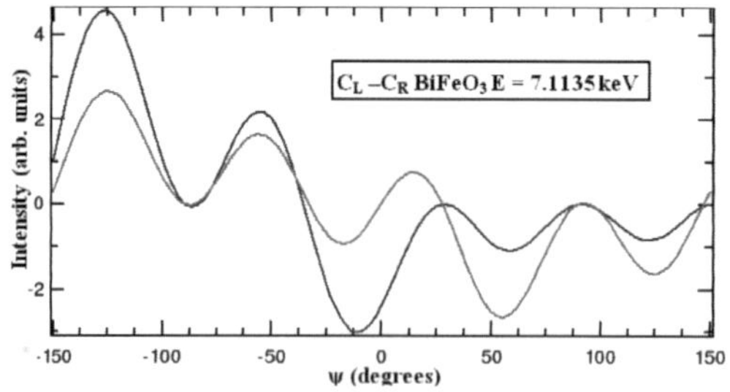

Figure 4. Simulation of the dichroic signal from Templeton-Templeton scattering under circular polarized x-rays for the reflection $(0,0,9)_H$ forbidden in R3c at the Fe K-edge. (Red line) Above and (blue line) below T_N (640K).

4. Conclusions

The study by X-ray resonant Bragg diffraction of the forbidden reflection $(009)_H$ for the R3c space group has revealed electronic properties of a single crystal of $BiFeO_3$. Existence of T&T scattering by quadrupoles in a cycloid implies a chemical structure that belongs to an enantiomorphic crystal class lacking a centre of symmetry. Space-group R3 (#146), one of 65 members of the Sohncke sub-group of crystal structures, is a maximal non-isomorphic subgroup of the nominal R3c-group, and thus a likely candidate for a commensurate chiral motif in bismuth ferrite. In this case, a chiral motif and a single domain are implied for the magnetically ordered state, and this does appear to be the case [25]. Quadrupoles (also higher-order multipoles) as a primary order-parameter is not unusual. However, again, order is achieved at low temperatures, because the underlying mechanism is weak [26-28]. The origin of the charge-like quadrupoles that contribute to T&T scattering could be related to bismuth 6s-6p lone pairs, known to drive certain structural distortions. Apart from expected direct hybridization of lone pairs, there is scope for admixture through the agency of oxygen 2p states that contribute to angular anisotropy in valence states observed at iron sites.

Acknowledgements

We acknowledge the Diamond Light Source for the beamtime allocation on I16. One of us (SWL) is grateful to Dr D D Khalyavin and Dr K S Knight for valuable discussion about the explanation of results offered in the communication. We are also grateful with G Catalan, who has provided the single crystal for performing the experiment. Financial support has been received from Spanish FEDER-MiCiNN Grant No. Mat2011-27573-C04-02. One of us (ARF) is grateful to Gobierno del Principado de Asturias for the financial support from Plan de Ciencia, Tecnología e innovación (PTCI) de Asturias. We thank Diamond Light Source for access to beamline I16 (MT-7720) that contributed to the results presented here.

References

[1] Lovesey S W, Knight K S and Balcar E 2007 *J. Phys.: Condens. Matter* **19** 376205
[2] Lovesey S W, Rodriguez-Fernandez A, Blanco J A 2011 *Phys.Rev.B* **83** 054427
[3] Lovesey S W, Fernández Rodríguez J, Blanco J A, Sivia D S, Knight K S and Paolasini L 007 *Phys.Rev.B* **75** 014409
[4] Catalan G and Scott J F 2009 *Advance Materials* **21** 2463–2485
[5] Kadomtseva A M, Zvezdin A K, Popov Y F, Pyatakov A P and Vorob'ev G P 2004 *J. Exp. Theor. Phys. Lett.* **79** 571
[6] Scott J F 2010 *Adv. Mater.* **22** 2106
[7] Eerenstein, W 2006 Nature **422** 759
[8] Lovesey S W, Balcar E, Knight K S and Fernández-Rodriguez J 2005 *Phys. Rep.* **411** 233
[9] Lovesey S W and Balcar E 2013 *J. Phys. Soc. Japan* **82** 021008
[10] Lovesey S W and Scagnoli V 2009 *J. Phys.: Condens. Matter* **21** 474214
[11] Wilkins S B, Caciuffo R, Detlefs C, Rebizant J, Colineau E, Wastin F and Lander G H 2006 *Phys. Rev. B* **73** 060406
[12] Fernandez-Rodriguez J, Lovesey S W and Blanco J A 2008 *Phys. Rev. B* **77** 094441
[13] Rodriguez-Fernandez A, Lovesey S W, Scagnoli V, Staub U, Walker H C, Shukla D K, Strempfer J and Blanco J A 2013 *Phys.Rev.B* **88** 094437
[14] Scagnoli V and Lovesey S W 2009 *Phys. Rev. B* **79** 035111
[15] Przeniosło R, Regulski M and Sosnowska I 2006 *J. Phys. Soc. Japan* **75** 084718
[16] Hannon J P, Trammell G T, Blume M and Gibbs D 1988 *Phys. Rev. Lett.* **61** 1245; ibid 1989 **62** 2644 (E)
[17] Lovesey S W and Balcar E 1997 *J. Phys.-Condens. Matter* **9** 8679
[18] Carra P, Thole B T, Altarelli M and Wang X 1993 *Phys. Rev. Lett.* **70** 694
[19] Lovesey S W 1998 *J. Phys.: Condens. Matter* **10** 2505
[20] Palewicz A, Sosnowska I, Przeniosło R and Hewat A 2010 *Acta Physica Polonica* A **117** 296
[21] Lebeugle D, Colson D, Forget A, Viret M, Bataille A M and Gukasov A 2008 *Phys. Rev. Lett.* **100** 227602
[22] Lee S, Choi T, Ratcliff W, Erwin R, Cheong S W and Kiryukhin V 2008 *Phys. Rev. B* **78** 100101
[23] Rodriguez-Fernandez A, Lovesey S W, Collins S P, Nisbet G and Blanco J A 2014 *J. Phys. Soc. Japan* **83**, 013706
[24] Templeton D H and Templeton L K 1985 Acta Crystallogr. **41** 365; ibid 1986 **42** 478
[25] Johnson R D, Barone P, Bombardi A, Bean R J, Picozzi S, Radaelli P G, Oh Y S, Cheong S W and Chapon L C 2013 *Phys. Rev. Lett.* **110** 217206
[26] Morin P, Schmitt D and de Lacheisserie E 1982 *J. Magn. Magn. Mater.* **30** 257
[27] Sakakibara T, Tayama T, Onimaru T, Aoki D, Onuki Y, Sugawara H, Aoki Y and Sato H 2003 *J. Phys.: Condens. Matter* **15** S2055
[28] Kuramoto Y, Kusunose H and Kiss A 2009 *J. Phys. Soc. Japan* **78** 072001

AUTHOR INDEX

Bérar, J-F 35
Beutier, G 20, 39, 66
Blackburn, E 66
Blanco, J A 80
Blasco, J 53
Bombardi, A 26, 66
Boudet, N 35
Caciuffo, R 66
Chapon, L C 26
Cheong, S-W 26
Collins, S P 20, 39, 80
Dmitrienko, V E 20, 39
Dudzik, E 49
Feyerherm, R 49
García, J 53
Grenier, S 1
Hayashi, K 35
Helliwell, J R 8
Herrero-Martín, J 53
Hosokawa, S 35
Igarashi, J-I 59
Islam, Z 49
Joly, Y 1
Kato, H 35
Lander, G H 66
Lee, N 26
Lim, J A 66
Livet, F 66
Lovesey, S W 80
Magnani, N 66
Meyer, D C 74
Nentwich, M 74
Nisbet, G 20, 39, 80
Ovchinnikova, E N 20, 39
Pilgrim, W-C 35
Radaelli, P G 26
Richter, C 74
Rodriguez-Fernandez, A 80
Sánchez, M C 53
Shukla, D K 49
Skaugen, A 49
Strempfer, J 49
Subías, G 53
Takahashi, M 59
Vecchini, C 26
Yamamoto, T 35
Zschornak, M 74